# The Architecture of Reform

## GEMS and National Standards

### Featuring "To Build A House"

by
**Dr. Cary I. Sneider**
**GEMS Science Curriculum Specialist**
*with* **Jacqueline Barber** and **Lincoln Bergman**

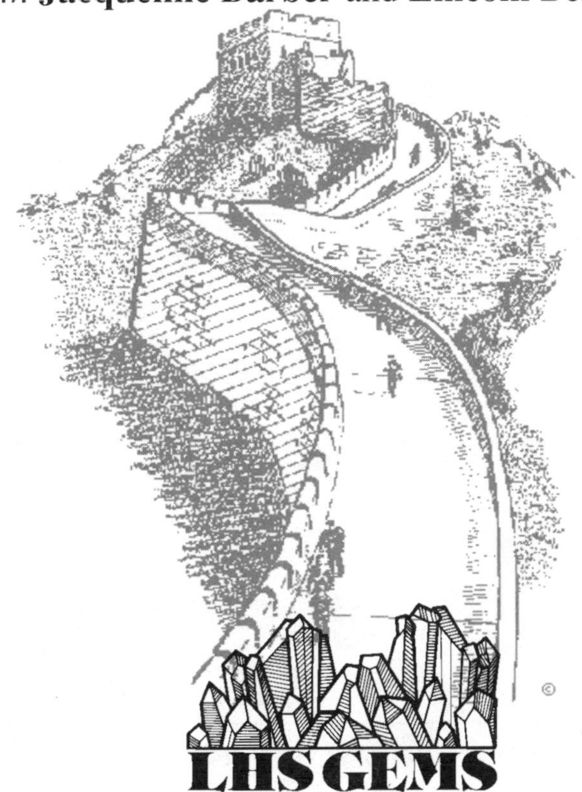

**Great Explorations in Math and Science**
**Lawrence Hall of Science**
**University of California**
**Berkeley, CA 94720-5200**

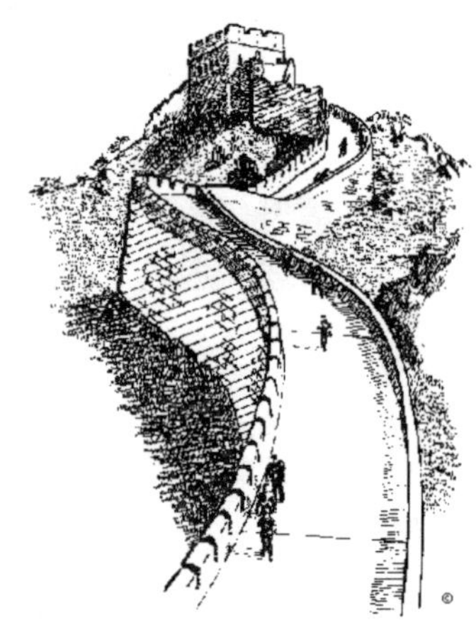

One of the wonders of the world, the Great Wall of China stands as a testament to the collective labor and creative power of human beings. Stretching for 4,000 miles across the northern border of China, the Great Wall, intended as a barrier to invading armies, also served as a road that facilitated travel. Similarly, the stack of new books on science education reform might be seen as a barrier by those who don't have the time to read them all, or they may be seen as a roadway, helping us find our way across rough terrain.

Lawrence Hall of Science
University of California,
Berkeley, CA 94720-5200

**Chairman:** Glenn T. Seaborg (in memoriam)
**Director:** Ian Carmichael

Initial support for the origination and publication of the GEMS series was provided by the A.W. Mellon Foundation and the Carnegie Corporation of New York. Under a grant from the National Science Foundation, GEMS Leader's Workshops have been held across the country. GEMS has also received support from: the McDonnell-Douglas Foundation and the McDonnell-Douglas Employee's Community Fund; the Hewlett Packard Company; the people at Chevron USA; Join Hands, the Health and Safety Educational Alliance; the Microscopy Society of America (MSA); and the Shell Oil Company Foundation. GEMS also gratefully acknowledges the contribution of word processing equipment from Apple Computer, Inc. This support does not imply responsibility for statements or views expressed in publications of the GEMS program. For further information on GEMS leadership opportunities, or to receive a catalog and the *GEMS Network News*, please contact GEMS at the address and phone number below. We also welcome letters to the *GEMS Network News*.



International Standard Book Number: 0-924886-37-4

---

### COMMENTS WELCOME !

Great Explorations in Math and Science (GEMS) is an ongoing curriculum development project. GEMS guides and handbooks are revised periodically, to incorporate teacher comments and new approaches. We welcome your criticisms, suggestions, helpful hints, and anecdotes about your experience presenting GEMS activities. Your suggestions will be reviewed each time a GEMS guide or handbook is revised. Please send your comments to: GEMS Revisions, c/o Lawrence Hall of Science, University of California, Berkeley, CA 94720-5200.
The phone number is (510) 642-7771.
The fax number is (510) 643-0309.

---

# Contents



---

## To the Reader

This *Handbook* represents an evolution of *To Build a House: GEMS and the Thematic Approach to Teaching Science* (1991), which was written before the publication of *Benchmarks for Science Literacy* (1993), the *Scope, Sequence, and Coordination Content Core* (1992), and the *National Science Education Standards* (1996), all of which are having a profound effect on science education in the United States. The GEMS staff hopes that *The Architecture of Reform* will enable all users of inquiry-based science activities to recognize the value of these important documents, and incorporate insights from them into their curriculum planning efforts. Most importantly, we trust this handbook clearly demonstrates how GEMS units can support the worthy goals of these documents in flexible, effective, and dynamic ways. We welcome your ideas!

# Great Explorations in Math and Science (GEMS) Program

The Lawrence Hall of Science (LHS) is a public science, teacher education, and curriculum development center on the University of California at Berkeley campus. Over the years, LHS staff have developed a multitude of activities, assembly programs, classes, and interactive exhibits. These programs have proven successful at the Hall and should be useful to schools, other science centers, museums, and community groups. A number of these guided-discovery activities have been published under the Great Explorations in Math and Science (GEMS) title, after an extensive refinement and adaptation process that includes classroom testing of trial versions, modifications to ensure the use of easy-to-obtain materials, with carefully written and edited step-by-step instructions and background information to allow presentation by teachers without special background in mathematics or science. The series now includes more than 50 teacher's guides and a number of handbooks on important topics, of which *The Architecture of Reform* is one. GEMS is also a nationally-recognized resource for professional development, offering a wide variety of inquiry-based workshops, with a growing national network of GEMS sites and centers

## GEMS Staff

**Principal Investigator:** Glenn T. Seaborg
**Director:** Jacqueline Barber
**Associate Director:** Kimi Hosoume
**Associate Director/Principal Editor:** Lincoln Bergman
**Science Curriculum Specialist:** Cary Sneider
**Mathematics Curriculum Specialist:** Jaine Kopp
**GEMS Network Director:** Carolyn Willard
**GEMS Workshop Coordinator:** Laura Tucker
**Staff Development Specialists:** Lynn Barakos, Katharine Barrett,
Kevin Beals, Ellen Blinderman, Beatrice Boffen, Gigi Dornfest,
John Erickson, Stan Fukunaga, Philip Gonsalves,
Linda Lipner, Debra Sutter
**Financial Assistant:** Alice Olivier
**Distribution Coordinator:** Karen Milligan
**Workshop Administrator:** Terry Cort
**Materials Manager:** Vivian Tong
**Distribution Representative:** Felicia Roston
**Shipping Assistants:** Ben Arreguy, Bryan Burd
**GEMS Marketing and Promotion Director:** Gerri Ginsburg
**Marketing Representative:** Matthew Osborn
**Senior Editor:** Carl Babcock
**Editor:** Florence Stone
**Principal Publications Coordinator:** Kay Fairwell
**Art Director:** Lisa Haderlie Baker
**Designers:** Carol Bevilacqua, Rose Craig, Lisa Klofkorn
**Staff Assistants:** Kasia Bukowinski, Larry Gates, Trina Huynh,
Steve Lim, Jim Orosco, Christine Tong

## Special Acknowledgments

The authors are particularly appreciative of the helpful reviews of this document by: Jo Ellen Roseman, Curriculum Director of Project 2061 of the American Association for the Advancement of Science; Elizabeth Stage, Project Director, New Standards Project; Herbert Thier, Director of the Science Education for Public Understanding Program (SEPUP) at the Lawrence Hall of Science, and Carolyn Willard, GEMS Network Director. Kimi Hosoume, GEMS Associate Director, contributed to the conception of this document as a major revision of the previous GEMS *To Build A House* handbook. While we endeavored to respond to issues raised by reviewers, final responsibility for accuracy and for statements or views in this handbook of course rests with the authors. Extensive work by Karen Ostlund, Professor of Science Education at Southwest Texas State University and GEMS Texas Network Coordinator, with GEMS Associates Mimi Halferty, Tracey Harros, and Pat Jones, helped us create the GEMS & National Standards chart in Appendix D.

# Introduction

Teachers and administrators who are responsible for planning school and district science programs find themselves in a world over-rich with guidelines on what the science curriculum should include, but impoverished in time and money to create real programs for students that reflect these guidelines in effective ways. The task is challenging, but by no means impossible. Charting a course through the maze of frameworks, benchmarks, and standards to get a handle on what science education reform is all about is the first step. Choosing a planning team and coming up with a practical plan for your school or district is the second. And, of course, turning the plan into classroom practice is the third and most challenging step. The urgent need to spread science and mathematics literacy to all students is the foundation stone for the transformations required as we enter the 21st century.

This handbook is designed to help you with all three stages of curriculum planning and implementation. Although it is produced by the GEMS Program at the Lawrence Hall of Science, we hope it will be useful in planning science programs that are consistent with state and national guidelines, no matter what science materials you intend to use, utilizing the many excellent programs now available or under development.

*The Architecture of Reform* does not propose new benchmarks or standards. Rather, it is intended as an accessible road map for people who want to understand and use the new documents to reform their own science education programs. It is also a demonstration, with numerous concrete examples, of how GEMS activities in particular and inquiry-based programs in general can contribute to bringing the common vision of all the reform into documents closer to reality. The book is organized as follows:

Part I is intended for readers who want to become familiar with the key documents in science education reform so that they can decide which documents they want to spend more time on and then move ahead in their planning process. Part I is written from the viewpoint of an imaginary writer who lives in the 21st century, when the reforms have successfully been implemented. The first chapter puts current initiatives into historical perspective, explaining how they connect to past efforts. Chapter Two describes the unique role of each of the major documents. Chapter Three discusses the common vision of all of the science education reform programs, and Chapter Four presents the idea that developing scientific literacy is much like building a "house of science."

Part II of this handbook relates the vision and goals of these key reform documents to the GEMS series. Chapter Five describes how the GEMS series supports the goals of the new reforms. Chapter Six discusses how sequences of GEMS units can be used as building blocks, either by themselves or in conjunction with other hands-on, minds-on programs, as enrichment experiences, or to create a curriculum for each grade level, from pre-Kindergarten through the first year of high school. These sequences are related to the *National Science Education Standards*. Chapter Seven provides suggestions for how to engage in local curriculum planning without having to reinvent the wheel. It also outlines systems that you will need to set up in order to implement and sustain an effective science program.

The appendices provide greater detail about some of the issues discussed in the text. Appendix A compares what each of the three major reform programs have to say about key ideas that are central to the current reform movement.

Appendix B defines ten *themes*, or *unifying concepts and processes* that help students connect different disciplines and topics in science.

Appendix C summarizes different aspects of the *nature of science* that are featured in GEMS Guides.

Appendix D is a chart providing a preliminary look at GEMS guides and National Standards criteria. [*Note*: The Lawrence Hall of Science (LHS) is currently engaged in a collaborative effort to correlate all curricula developed by LHS, including FOSS, SEPUP, GEMS, and many others, to the *National Science Education Standards*. That correlation document will provide detailed discussion, including concrete examples and direct quotations from the various curricula and the standards, to better enable teachers and educators to understand the ways LHS curricula embody scientific inquiry, convey "unifying concepts and processes," and bring other key aspects of the standards to students.]

Appendix E includes transparency masters based on the "House of Science" metaphor "constructed" in Chapter 4 that you can use to spark an informative and entertaining discussion at your school or district about science education reform. An earlier version of "To Build A House" was presented by many educators nationwide to illustrate the thematic approach to teaching science.

The world of science and mathematics education, like the subject matter from which it springs, is in a constant state of ferment and change. Truths and truisms that we took for granted a decade ago may not seem to self-evident today. As national standards and related guidelines and frameworks are discussed, debated, and, most importantly, implemented, more will be learned. Within this context of change, this handbook represents our attempt to synthesize the major building blocks of the leading reform documents and to summarize how well GEMS activities fit into that larger picture.

As Part One makes clear, we've discovered that the leading documents share a common vision, especially when viewed from the year

# 2020!

# Part I
## 2020 VISION

This section of *The Architecture of Reform* is intended to provide basic background and quick familiarity with the key documents to encourage deeper investigation and set the stage for curriculum planning.  This entire section is written from the viewpoint of an imaginary writer who lives in the 21st century, when the reforms have been successfully implemented.

**Chapter 1**    puts current initiatives into historical perspective, explaining how they connect to past efforts.

**Chapter 2**    describes the unique role of each of the major documents.

**Chapter 3**    discusses the common vision of all of the science education reform projects.

**Chapter 4**    presents the idea that constructing science literacy  is much like building a "house of science."

# Chapter 1

# The View from the 21st Century

## A Hundred Years at a Glance: 2020 Vision

As I sit writing this, in the year 2020, I am proud to say that nearly everyone in our nation is a scientifically literate citizen. As most readers know, it has not always been so. For most of the previous century, students learned science primarily from books. Learning was confined to the memorization of facts, and assessment consisted of multiple-choice testing. Laboratory sessions were aimed at confirming what the textbook said was true, and if students failed to get the right answers, they received poor grades.

The seeds of science education reform were sown early in the century by visionary reformers who wanted to place student-initiated inquiry and group activities at the center of the school experience. According to an article by Richard Elmore (1996), a wide variety of innovative experiments in education were undertaken in the early decades of the 20th century. In the late 1920s and early 1930s, a number of cities, including Denver, Colorado, and Washington, D.C., provided time for teachers to meet during the day to revise the curriculum and develop new teaching practices. Unfortunately, these early efforts lost steam by the 1940s, when critics attacked what they called "watered-down content" and a "focus on children's psychological adjustment at the expense of learning."

New energy was injected into the science education reform effort when Sputnik was launched by the Soviets in 1957.  Concerned that the United States was falling behind the Russians in science and technology, Congress appropriated large sums of money to develop new science curricula in which students could engage in inquiry—both to make discoveries about the natural world, and to gain firsthand knowledge about the thought processes and methods of inquiry used by practicing scientists.

The first new instructional materials for the high school level were developed during the 1960s with support from the National Science Foundation.  The new courses included PSSC Physics, BSCS Biology, CHEM Study, and Man, A Course of Study (MACOS).  The high school initiative was quickly followed by development of programs for the elementary level, including Science A Process Approach (SAPA), the Science Curriculum Improvement Study (SCIS), and Elementary Science Study (ESS).  Programs for the intermediate level were also created, including the Intermediate Science Curriculum Study (ISCS), and Introductory Physical Science (IPS).  In the 1970s various programs were developed to correct some of the shortcomings of the earlier courses.  Harvard Project Physics, for example, replaced the more rigorous and difficult PSSC course with a program aimed at a broader base of students, using a strong historical focus to illustrate the development of ideas, and laboratory activities to further clarify important concepts.

The vision of good science education of that period was similar to that of the early reformers, and the kind of science teaching that most of us value today.  Students were encouraged to explore scientific concepts, to see how scientific principles and theories apply in the world around them, and to work with other students in conducting investigations.  Through summer institutes sponsored by the National Science Foundation, hundreds of thousands of teachers were trained in the use of the new materials, and millions of students were exposed to hands-on activities.

Unfortunately, funding to the National Science Foundation for education was cut to the bone in the mid-1970s.  With fewer opportunities to train teachers in the new methods, and reduction in the flow of new instructional materials, most schools soon returned to textbook-driven teaching in which students were asked to memorize terms, do cook-book labs, and respond on multiple-choice tests.  Unfortunately, this situation existed in most schools until well into the 1990s (Stake and Easley, 1978, Elmore, 1993, NRC 1996).

In Elmore's view, an important reason that all of the past efforts to reform science education have never reached a majority of schools is that there have never been sufficient incentives to change.  Providing good instructional materials and training for some teachers in how to use them is not enough.  Continuing support and incentives need to be provided by school administrators and community members, as well as from professional organizations, universities, and credentialing boards.  Longterm, consistent, sustained support and follow-up is needed.  Parent, caregiver, and community involvement and understanding are crucial.  There also needs to be a clear vision and consensus concerning what constitutes good science education.

> *The educational foundations of our society are presently being eroded by a rising tide of mediocrity that threatens our very future as a Nation and a people.*
> —A Nation At Risk  (1983, page 5)

The call for reform was renewed in 1983, when the National Commission on Excellence in Education issued its report, *A Nation At Risk*.  The report described what has happened to our system of education in very strong terms: "If an unfriendly foreign power had attempted to impose on America the mediocre educational performance that exists today, we might have viewed it as an act of war.  As it stands, we have allowed this to happen to ourselves.  We have even squandered the gains in student achievement made in the wake of the Sputnik challenge.  Moreover, we have dismantled essential support systems which helped make those gains possible.  We have, in effect, been committing an act of unthinking, unilateral disarmament." (*A Nation At Risk*, page 5)

*A Nation At Risk* painted a picture of our science education establishment as obsolete and in ruins.  Using architecture as a source of metaphors, an apt metaphor for science education in the United States as depicted by *A Nation at Risk* might be the Egyptian pyramids or an ancient Greek temple.   The ruins of these magnificent structures speak to us over the centuries of the decline and fall of ancient civilizations. It was such a decline that the authors of A Nation at Risk warned against.   Because of its strong language, and the integrity and accomplishments of its authors, *A Nation at Risk* served as a very effective catalyst for change.

The authors of *A Nation at Risk* did not simply step aside after their report was published, and expect others to do something to improve our science and mathematics system of education. One of the authors was Dr. Glenn T. Seaborg, Nobel Laureate in Chemistry and Chairman of the Lawrence Hall of Science. Dr. Seaborg actively encouraged development of a number of new curriculum projects and programs at the Lawrence Hall of Science that would respond to the realities of the educational system, while establishing new directions for the 21st century. One of those was the Great Explorations in Math and Science (GEMS) program, which received its first grants from the A.W. Mellon Foundation and the Carnegie Corporation of New York, in 1985. Dr. Seaborg served as Principal Investigator for the GEMS program.

The GEMS program (and many others) responded to the realities of the period. Science and mathematics education generally commanded a small portion of state and federal budgets, and very few elementary and middle school teachers had formal training in science. GEMS Teacher's Guides were written so that a teacher who has no formal science background can present an excellent inquiry-based program at low cost. The materials proved to be very popular with teachers. Not only did students enjoy doing the fun and challenging activities— they also learned pivotal science concepts and improved their inquiry skills.

Now, in the year 2020, GEMS is still a popular science program, although today more funds are available for science and more teachers have strong science backgrounds. Many teachers recall that their careers were launched when *they* were involved in GEMS activities in elementary and middle school. Some teachers still use updated versions of those early classics *Oobleck* and *Bubble-ology*, even though many new GEMS Teacher's Guides are now available.

Naturally, GEMS was not the only program that took hold at the beginning of the new millennium. Many other programs for all age levels were developed at the Lawrence Hall of Science, including the Full Option Science System (FOSS) which was adopted by many states and districts nationwide and the Science Education for Public Understanding Program (SEPUP) which brought activities into middle and high school classrooms that focused on science and societal issues. In other parts of the country, other organizations and educators developed and began publishing excellent curricula. A critical mass of teachers and schools adopted a more active, inquiry-based science program.

With historical hindsight, it is not difficult to see how these new activity-based science programs contributed greatly to the movement towards nationwide science education reform. For one thing, they were built upon a solid foundation of research and earlier experimentation. Hands-on minds-on learning approaches were not new. The seeds of progressive science education were sown by John Dewey and others at the beginning of the 20th century. These early ideas were bolstered later in the century by the work of cognitive psychologists such as Jean Piaget, adding important insights into child development and the teaching-learning process. These insights were in turn built upon by educators such as Robert Karplus, whose learning cycle concept helped teachers and curriculum developers recognize the different steps of the inquiry process, and design a great many effective activity-based lessons. As the 20th century came to a close, there was widespread agreement among educators that an inquiry-based, experiential, constructivist approach which took into account the different ideas that students brought into the classroom was key to effective and lasting science education.

While these advances kept the inquiry approach alive and in the forefront of virtually all efforts to improve science education, it was not until century's end that the reforms took hold on a massive scale.  It was only then that the availability of the new materials, and growing opportunities for teacher education, coincided with support from above.  Stirred to action by *A Nation At Risk*, state and national policy-makers were at last ready to provide the support and incentives necessary to achieve broad-based change.

Many State Departments of Education developed guidelines, often called "frameworks," to specify the nature and practice of good science education.  In addition to these state-based efforts, several national organizations undertook projects to chart a course towards reform.  These organizations included the American Association for the Advancement of Science (AAAS), the National Science Teacher's Association (NSTA), and the National Research Council (NRC), the operating arm of the National Academy of Sciences.

To a very large extent, we owe a debt of gratitude to these state and national policy-makers for their common vision of a scientifically literate citizenry.  That vision has led to the establishment of inquiry-based programs in virtually every school in our nation.  And the students who have been involved in these programs have significantly contributed to the prosperity and security of our nation in the opening decades of the 21st century—helping to move humanity towards a more sustainable future.

It's not surprising that when the initial reform documents came out during the early 1990s, teachers and administrators were overwhelmed.  Given the task of revamping their district science program, it was not uncommon to hear science committee members wonder which, if any of the documents, they should use as a guide.  With our 2020 hindsight, it is not difficult to see that all of these documents supported a certain core of ideas, and each made a unique contribution to local plans and programs.  The next chapter describes the three major national projects that raised the reform effort to a higher level, and the unique value of each one in helping local schools and districts bring about the needed reforms.

---

## A Special Mathematical Note

In the narrative that follows, our commentator from the year 2020 is focusing on documents relating to reform in *science* education.  It is important to note, however, that mathematics educators and their organizations have often provided primary impetus for transforming ideas and new directions.  Standards from the National Council of Teachers of Mathematics (NCTM) for example, were among the first to place the words "national" and "curriculum" in close proximity, thus breaking a previous "taboo" and helping pave the way for other concepts of national scope, such as national standards, to be taken seriously.  Another new GEMS handbook, *The Rainbow of Mathematics*, is now under development, which will discuss in much greater detail the NCTM Standards, the major math strands, and how GEMS activities can be used as a strong support to these leading criteria for mathematical educational excellence.

# Chapter 2

# The Reform Documents

From our 21st century vantage point, we can see that each of the major documents created during the closing decades of the 20th century played an important and complementary role in making our science education system what it is today.  In this section, I'll use architecture as a useful metaphor for visualizing the unique role played by each document, and helping us to keep these complementary roles in mind.

**State Science Frameworks.**  When the critical reports hit the headlines in the 1980s, a few State Departments of Education had already begun to develop guidelines for good educational practices in all areas of the curriculum.  The state of California, for example, issued a *Science Framework for California Public Schools* in 1978, and a *Science Framework Addendum* in 1984, which set the goal of achieving science literacy for all students.  Other states produced frameworks and other sets of guidelines, and the revision and publication of these documents continues today.

A useful metaphor for these state guidelines might be the steel I-beam frameworks built to support large public buildings. Recognizing that local control of schools is necessary, desirable, and a fact of life in the U.S., those who constructed the state frameworks intended to provide a flexible supporting structure, which would be adapted, modified, and filled in by individual schools and districts.

Until the critical reports of the 1980s, however, no major attempt was made to influence national science education policy.  Then, several national organizations sought to clarify the vision.

**Project 2061.**  The American Association for the Advancement of Science (AAAS) is the home of  Project 2061, a long-range effort designed to help the nation achieve scientific literacy.  The project was started in 1985, when our planet was visited by Halley's Comet.  It was named for the year 2061, when children who were then just starting school could expect to see the comet return to the inner solar system.  The project aimed to create Reform Tools to help schools and school districts assemble their own science education programs.

*Science for All Americans* (1989) was the first Reform Tool created by Project 2061. In contrast to the earlier reports of gloom and doom, *Science for All Americans* projected an articulate and optimistic vision of a scientifically literate populace.  By describing what students should know and be able to do when they graduate from 12th grade, *Science for All Americans* provided an operational definition of scientific literacy. Created  by the work of scientists who had been asked to ignore previous textbooks in deciding what is important for everyone to know in the 21st century, this report "broke the mold," and greatly expanded the vision of what science programs should include.  In addition to the natural sciences, chapters included The Nature of Science, Mathematics, and Technology, Human Society, The Designed World, Historical Perspectives, Common Themes, and Habits of Mind.  The soaring and beautifully designed Taj Majal would be a good metaphor for the elegant vision of science education projected by *Science for All Americans*.

Although it projects an exciting vision for scientific literacy, *Science for All Americans*  did not provide specific guidance to teachers of lower grades.  It was difficult for a first or second grade teacher to translate what students should know when they graduate from 12th grade into useful guidelines for planning the year's work.  Something else was needed.

*Benchmarks for Science Literacy*  (1993) was produced by Project 2061 in response to the call for a more explicit set of guidelines for what students should know and be able to do by the end of 2nd, 5th, 8th, and 12th grades.  In all, more than 850 individual

U.S. Science Education as viewed in  ***Science for All Americans*** **(1990)**

benchmarks were identified, organized by chapter heading and by grade level within each chapter.  The benchmarks were elaborated with narrative descriptions of the kinds of activities that will lead to understanding, and descriptions of research on teaching these concepts.  There was a very careful effort to remove jargon and make the statements as practical as possible.

Because of its clear, layered structure, consisting of many individual benchmarks assembled into a whole, the *Benchmarks for Science Literacy* could be represented as the Eiffel Tower.  The clarity of the *Benchmarks*, and the careful selection of concepts that corresponded to stages of cognitive development, led to its becoming a very useful document for educators and teachers who were responsible for selecting science content to be taught at each grade level in their schools or districts.

Project 2061 also released *Resources for Science Literacy: Professional Development*, which is a CD-ROM and companion print volume for use by science educators.  The CD-ROM contained the text of *Science for All Americans* with connections to various databases, including: citations of 120 Science Trade Books, descriptions of 15 College Courses, detailed Comparisons of Benchmarks and National Standards, a Cognitive Research Database, and a Project 2061 Workshop Guide.

Among other Reform Tools proposed by Project 2061 were: *Designs for Science Literacy*, a handbook for educators wishing to take a systematic, goal-oriented approach to curriculum design; *Atlas of Science Literacy*, a collection of growth-of-understanding maps that graphically portray the sequence and interdependence of knowledge and skills that lead to scientific literacy; and *Blueprints for Reform*, a collection of papers by experts on aspects of the system that must change in order to accommodate the reforms being proposed by Project 2061.

***Benchmarks for Science Literacy***

***Table of Contents***

1. Nature of Science

2. Nature of Mathematics

3. Nature of Technology

4. Physical Setting

5. Living Environment

6. Human Organism

7. Human Society

8. Designed World

9. Mathematical World

10. Historical Perspectives

11. Common Themes

12. Habits of Mind

**The Scope, Sequence, and Coordination (SS&C) Project.**  The National Science Teacher's Association (NSTA) created the Scope, Sequence, and Coordination (SS&C) Project in 1989, to restructure science education at the 6th-12th grade levels.  The project intended to replace the "layer cake" model, in which high school students study one science discipline each year, with a new model, in which every student studies every science every year, from grade six through grade twelve.  The project provided specific guidelines for creating new instructional materials, and for selecting materials to plan the school science curriculum.  In brief, SS&C  advocated:

*Every science, every year, for every student, grades 6-12*

*Scope*: Follow the "less is more" principle in deciding what to cover.  According to this principle, it is better for students to study fewer topics in depth, than to cover a large number of topics superficially.

*Sequence*: Design the curriculum so that students encounter concepts at successively higher levels of abstraction over several years, and have opportunities to apply them to problems of personal concern in the early years, and global issues in the later years.

*Coordination*: Plan the science program so that students see the connections among concepts in biology, chemistry, physics, and Earth/space sciences.

Unlike Project 2061, which was aimed at replacing the entire body of science content, the SS&C project valued much of what had previously been included in science, but wanted to reorganize it so that students could see the connections among the disciplines, revisit key concepts at successively higher levels, and learn science in a way that is developmentally appropriate, and related to personal concerns and global issues.  While the SS & C Project was eventually superseded and incorporated by the *National Science Education Standards*, its important historical role in providing guidelines for and restructuring the middle and high school science program should be acknowledged.

To represent the idea that certain elements of the previous system should be incorporated in a new and elegant design, we have used the metaphor of the Nation's Capitol, which includes certain elements of ancient Greek and Roman architecture (tall marble columns, triangular pediments, arched windows, and domes) reorganized into a new and coherent vision.

U.S. Science Education as viewed in *Scope, Sequence, & Coordination* **(1992)**

*__National Science Education Standards.__* Although Project 2061 and the SS&C project set the tone for science education reform, by the early 1990s, policy-makers felt that more was needed.  In order for the country to move ahead, a national consensus was called for that would embrace science content, teaching, and assessment.  Enter the most prestigious organization of scientists in the United States—the National Academy of Sciences.

The National Research Council (NRC) is the operating agency of the National Academy of Sciences, and is frequently called upon by Congress or other government agencies to create definitive reports on issues related to science and technology.  In 1992, the NRC established a committee to create standards for science content, teaching, and assessment.  In order to achieve a national consensus on what the standards should contain, copies of a draft document were distributed to 18,000 individuals and 250 groups.  Comments were analyzed and incorporated into the final draft of the *National Science Education Standards* which was released in January, 1996.

The *National Science Education Standards* (1996) might be depicted as the entire City of New York, with its vast and culturally diverse population, its new buildings, inspirational monuments and means of transportation and food distribution systems, and so on.  In short, the vision projected by the *National Science Education Standards* is a complete and comprehensive vision of how the educational system needs to change in order to  give every student an opportunity to become scientifically literate.

### *The National Science Education Standards* (1996)

## Table of Contents

1. Introduction
2. Principles and Definitions
3. Science Teaching Standards
4. Professional Development Standards for Teachers
5. Assessment Standards
6. Science Content Standards

   K-12 Unifying Concepts and Processes

K-4
5-8
9-12

Science as Inquiry
Physical Science
Life Science
Earth and Space Science
Science & Technology
Science in Personal & Social Perspectives
History and Nature of Science

7. Program Standards
8. System Standards

The largest single chapter in the *National Science Education Standards* presents a list of what students at all grade levels should know and be able to do. Within a few months of the release of the *Standards*, several articles had been written showing that the content of this chapter was similar, in many ways, to the content in *Benchmarks for Science Literacy*. Further, in the opening pages it identifies the *Benchmarks* as entirely consistent with *National Science Education Standards*.

So, the unique role played by the *National Science Education Standards* was not to provide an image of a scientifically literate person, since that had already been done by Project 2061. Instead, it was to project the most coherent vision ever of what should occur in the science classroom, and of how teachers should be educated in order to create excellent programs for their students. In addition, the *National Standards* discussed alternative methods of assessment and how assessment data can be used to improve science education. Finally, in chapters on program standards and system standards, the document discussed in very broad terms how to evaluate local science programs, and how society as a whole can and should contribute to science education reform. This comprehensive and systemic orientation towards improving science education makes a compelling argument for mobilizing the necessary government and community resources to change the educational system as a whole.

In summary, each of the projects and documents discussed in this chapter played a very important role in science education reform:

- **State Science Frameworks** drove the reform effort in each state.

- *Science for All Americans* provided an operating definition of scientific literacy.

- *Benchmarks for Science Literacy* provided a clear and comprehensive listing of what every student should know and be able to do by the end of the 3rd, 5th, 8th, and 12th grade levels.

- *SS&C Content Core* established guidelines for restructuring the school science experience for students in grades 6 through 12.

- *The National Standards for Science Education* provided a complete and comprehensive vision for science education and the systems that are needed to support it.

While each project played a distinct role, together these documents projected a common vision. In the end, it was this commonly held vision that came through, and helped to create our 21st century system of science education. The common vision of the reform projects is the subject of the next chapter.

---

*Reports and recommendations from the **Third International Math and Science Study (TIMSS)** have also been influential in focusing attention on the need for major reforms (as well as providing international comparisons). Especially important toward the end of 20th century was the idea that U.S students, as compared to other leading nations, received instruction, often via textbooks, in a very **wide** selection of topics, but all too often the result was superficial coverage and limited understanding. This was sometimes expressed as "a mile wide and an inch deep," and this phrase became a clarion call for the new reformers, adding weight to the "less is more" idea and emphasizing the goal of teaching fewer units in greater depth.*

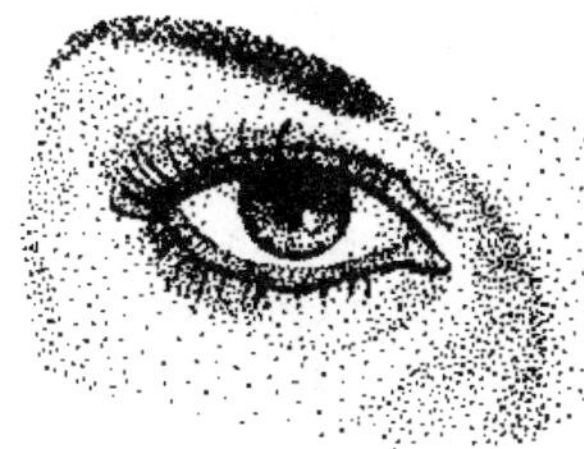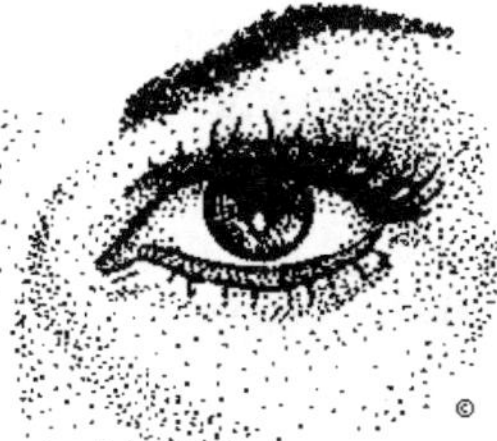

# Chapter 3
# A Common Vision

Leafing through the documents that set the stage for 21st century science education, their most striking quality is not their differences, but the common vision that they projected about the fundamental purpose of science education. They were in essential agreement on questions such as: who should study science, what students should learn, how they should learn it, and what must happen to implement the vision. This common vision will conclude my report, written just as the new school year is about to start in September, 2020.

## Science for Whom?

In the common vision of the new reform movement, *every student* should have an opportunity to become a scientifically literate individual. This is a distinctly different goal than the one that drove science education reform in the 1950s and 1960s. When the Soviet spacecraft Sputnik soared into the heavens, many thought that our country was falling behind in scientific and technical expertise, so the initial focus was on rigorous high school science courses. Over time, fewer and fewer students enrolled in high school science courses, so that by 1989, only 25% of the nation's high school students took any science at all after the tenth grade biology required in most states (*SS&C Content Core*, page 14). In contrast, the goal that emerged by the end of the 20th century was that *all citizens* should have an opportunity to become scientifically literate. Furthermore, each of the guiding documents emphasized the inclusion of groups who have been historically underrepresented in scientific and technical fields, including girls and women, students from diverse racial, cultural, and economic backgrounds, and students with physical and learning challenges.

In summary, the answer to the question, "Science for Whom?" encompasses two fundamental ideas.

> 1) Science is for everyone; and
>
> 2) A commitment to diversity and equity, so that science educators reach out to groups who have historically been excluded from science and technical fields.

## What Should Students Learn?

In previous decades the idea of "hands-on" activities had been stressed. In the 1990s, this idea was carried further to clarify that use of the hands was not enough. "Hands-on and minds-on" was sometimes used to express the idea. The object of these activities was that students would learn inquiry skills. As expressed by the *National Standards*, "inquiry is a step beyond 'science as a process,' in which students learn skills such as observation, inference, and experimentation. The new vision includes the 'processes of science' and requires that students combine processes and scientific knowledge as they use scientific reasoning and critical thinking to develop their understanding of science." (*NSES*, page 1-5)

If we think of the pendulum of educational thought swinging back and forth between "content" and "process" every decade or so, the above statement places the *National Standards* (and other reform projects of the 1990s) squarely in the middle, where students are expected to learn both process skills *and* knowledge, along with the reasoning abilities needed to bring them together.

One of the causes of the crisis in science education was that until the last decade of the 20th century, the school science program had grown by accretion. As science advanced more and more rapidly, more and more topics were added to the curriculum. The problem was that nothing was taken away. As stated in *Science for All Americans* (page viii), the curriculum had become "overstuffed and undernourished." And as stated in the *National Science Education Standards:* "If teachers are to teach for understanding as described in the content standards, then coverage of great amounts of trivial, unconnected information must be eliminated from the curriculum." (*NSES*, page 213) The *SS&C Content Core* emphasized the same idea by stating the "less is more" principle, meaning that it is better for students to spend more time studying a smaller number of topics in-depth than to try and cover too many topics superficially.

However, deciding what to toss out and what to include was not a trivial task. One of the most important reasons that the science education reform movement succeeded was that a consensus was eventually achieved concerning what students should learn from the school science program. A few key ideas guided this very important effort.

One important idea is to provide a conceptual framework that will enable students to see the connections between different topics. In 1989, *Science for All Americans* was the first of the national reform documents to point out that certain overarching *themes*, or big ideas, guided research and theory in the sciences. Ideas such as patterns of change, scale and structure, evolution, and energy were used by scientists who worked in many different disciplines, to make sense of the natural world. It seemed reasonable that these ideas would also help students see the connections among the many disciplines and topics in their science courses. This idea was developed further in the *SS&C Content Core*, which proposed the development of high school courses based on *great ideas* such as energy or evolution, and by the *National Science Education Standards,* which started an important chapter on content with a section on *unifying concepts and processes*, including: systems, order and organization; evidence, models, and explanation; change, constancy, and measurement; evolution and equilibrium; form and function. We'll take a closer look at unifying concepts and processes in Appendix B.

A second idea is to challenge scientists to select only the most important concepts and theories that should be part of every person's education. Project 2061's *Benchmarks for Science Literacy,* and the *National Science Education Standards* provided clear and detailed descriptions of the most important concepts, principles, and theories for inclusion in the school science program. For the most part, these lists are in substantial agreement with each other (Project 2061 Newsletter; *NSES,* page 15). One analysis revealed a 90% agreement between the content chapters in the *Benchmarks* and the *National Standards* (Brearton, 1996).

Recognizing that scientists rarely confine their efforts to a single discipline in solving real world problems, the reform projects urged curriculum developers and teachers to look for new ways to organize science education. Instead of a course entitled Chemistry, for example, students might study Forensic Science, in which they combine chemistry with logic and inference to solve mysteries. Biology might be replaced with courses entitled Environmental Change, Ocean Ecology, or the Evolution of Life and Climate. Physics might be replaced by interdisciplinary courses in Space Science, Engineering, or Energy.

Two other ideas about what students should learn were also introduced in *Science for All Americans,* and reflected by the *SS&C Project* and *National Standards.* One is that in addition to the concepts and theories that students learn in the natural sciences, they should also learn about the nature and history of science. The other is that they should learn about societal issues that are related to science and technology. (This is sometimes called a science-technology-society, or STS approach.) In summary, several solutions to the "overstuffed and undernourished science curriculum" are to:

3) Start with a foundation of inquiry skills.

4) Connect concepts and theories within a framework of unifying principles.

5) Identify the most important concepts and theories that must be included in the curriculum.

6) Orient science around interesting topics that integrate the various fields of science.

7) Be sure to include the history and nature of science.

8) Present societal issues that are related to science and technology.

In one sense, these guidelines represent a narrowing of the content of school science to just the most important ideas. On the other hand, it represents an expansion of the scope of science, to include the history and nature of science, as well as relevant societal issues. *The National Standards* cautions that the aim of including these new areas is not to transform science education into social studies; but to enable students to understand the human side of science.

## How Should Students Be Taught?

Another important point of agreement grew from a century of research on how people learn. The term *constructivism* encompasses the recognition that students enter the classroom with already-formed ideas about the natural world, and that learning involves students in critically examining their initial intuitive ideas so as to construct new concepts which are more powerful and accurate ways of viewing the world. In other words, learning cannot be achieved by lecture and reading alone. Students need to have opportunities to struggle with new ideas, and to discover how they might agree or conflict with their current ideas before significant new learning can take place. The awareness that all students should have opportunities and experiences to assist them in constructing the concepts and theories of science transforms the essential conception of teachers into facilitators of learning, rather than as sources of knowledge alone. This idea was expressed in *Science for All Americans* as follows:

> *Students come to school with their own ideas, some correct and some not, about almost every topic they are likely to encounter. If their intuition and misconceptions are ignored or dismissed out of hand, their original beliefs are likely to win out in the long run, even though they may give the test answers their teachers want... students must be encouraged to develop new views by seeing how such views help them make better sense of the world.* (SFAA, page 186)

One important idea that emerged from the work of Dewey and others at the beginning of the twentieth century and survived to become a cornerstone of the new vision at century's end was the value of group work, the educational effectiveness and social benefit of cooperation and collaboration. It is also an accurate representation of the importance of collaboration in science. Specifically, students should spend a significant portion of time working with other students to conduct experiments, to interpret data, and passionately argue about ideas. This strategy reflected the real world of work in which teams of people work together on a myriad of socially constructive and needed tasks and projects. As expressed in the *National Science Education Standards:*

> *Working collaboratively with others not only enhances the understanding of science, it also fosters the practice of many of the skills, attitudes, and values that characterize science. Effective teachers design many of the activities for learning science to require group work, not simply as an exercise, but as essential to the inquiry. The teacher's role is to structure the groups and to teach students the skills that are needed to work together.* (NSES, page 50)

In Summary, two important ideas about teaching are that:

> 9) Students need to construct their own understanding of science; and
>
> 10) Students need to collaborate in teams much of the time.

**What Must Occur to Implement the New Vision?**

For many years, science educators were puzzled about why the "new" ideas about what to teach and how to teach never took hold and had to be reinvented in successive decades. Although there were some unique aspects in the vision of science education of the 1990s, most of the ideas about what and how to teach had been proposed decades before. As mentioned above, a consensus about what to teach and how to teach were very important factors. Another very important factor was agreement on how to assess student progress. For example, once teachers recognized that their students would not just be tested on information about experiments, but on their abilities to actually design and carry out an experiment, they changed their teaching methods to give their students opportunities to practice those capabilities. The *National Science Education Standards* gave great emphasis to this idea that: "...the alignment of assessment with curriculum and teaching is one of the most critical pieces of science education reform. If the assessment system at the school and district levels does not reflect the Standards and measure what is valued, the likelihood of reform is greatly diminished." (*NSES*, page 211)

Another unique quality of the shared vision of the 1990s reforms is the need for *systemic change*. Although viewed slightly differently by the three projects, all agreed that deep-seated systemic change was needed for the reforms to occur. The "system" included not only the instructional materials that were used, and the ways these were presented by the teacher in the classroom, but facets of the system that provided opportunities for ongoing teacher education, for student assessment, for teacher evaluation and rewards, and for involving the entire community in the educational process.

In summary, in order to implement the new reforms, it is necessary to:

---

11) Align assessment with curriculum and teaching; and

12) Mobilize the entire educational system to support the new reforms.

---

**The Common Vision**

The common vision of science education reform in the 1990s can be summarized in twelve key recommendations, listed on the following page. Although these are far from the only ideas expressed in the various reform documents, they are among the most important ideas that are common to all of them.

Of course, a great deal of the richness of the documents under discussion is lost in a simple summary. Each of the reform projects had somewhat different emphases and approaches. In Appendix A are twelve tables that will allow you to compare and contrast these different perspectives. Browsing through them will enrich and deepen your understanding of the fundamental ideas that led the science education community into the 21st century.

# A Common Vision

## for Science Education Reform in the 1990s

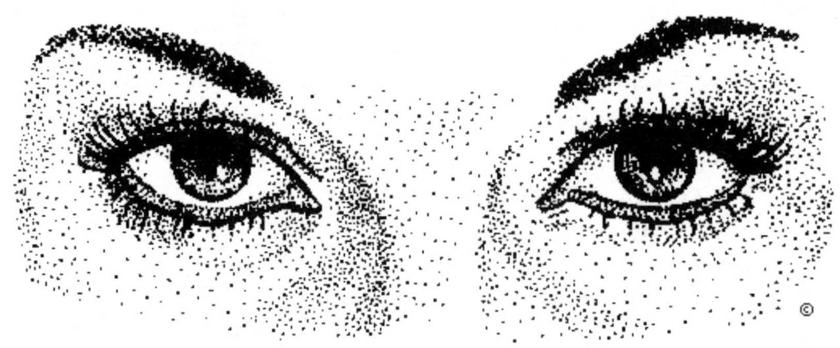

### Science for Whom?

1. Science is for *all* students.

2. Strong commitment to diversity and equity.

### What Should Students Learn?

3. Start with a foundation of inquiry skills.

4  Construct a framework of unifying concepts and processes.

5. Identify the most important concepts and theories.

6. Integrate ideas from different fields whenever possible.

7. Include examples illustrating the history and nature of science.

8. Show how science and technology inform societal issues.

### How Should Students be Taught?

9.  Students need to construct their own understandings of science.

10. Students need to collaborate in teams much of the time.

### What Must be Done to Implement the Vision?

11. Align assessment with curriculum and teaching.

12. Mobilize the whole system to support the new reforms.

# Chapter 4
# Building a House of Science

This chapter explores the metaphor of a "house of science" to represent the knowledge, abilities, and attitudes that students develop as they become scientifically literate individuals. Just as a house provides warmth, shelter from the elements, and a sense of place in the world, a "house of science" is a very important part of what every individual needs to participate fully in modern society.

In helping your students construct their houses of science, you will play many roles to facilitate learning. They will do their own construction—you can help provide your students with the experiences, challenges, materials, tools, and ideas that they will need to construct sturdy homes to support and nourish them through long and productive lives. In short, you can help them grow into scientifically literate men and women.

The metaphor begins by considering what it is that we want to teach our students, and how we can help them put these facts, concepts, and theories together into a coherent structure.

*During the late 1980s and 1990s, as the leading reform documents were being written, a new facet was added to the GEMS series—GEMS handbooks. Beginning as handbooks for specialized audiences (teachers, leaders, parents) these handbooks burgeoned into accessible treatises on key educational topics, including assessment, literature connections, and* To Build A House: GEMS and the Thematic Approach to Teaching Science. *The* Architecture of Reform *replaces and updates* To Build A House, *but we didn't want to lose the metaphor! The house-building comparison provided a clear and accessible framework within which to place what were then called "themes," (now called "essential concepts and processes" in the National Standards). The overheads provided for the "To Build A House" presentation have been used by educators nationwide as a compelling way to represent major reform ideas. In the following pages, we adapt this presentation to include the* National Standards *and other current thinking, as well as to represent, as it did before, our understanding of how teachers can help students construct their own unique houses of science.*

*Science is constructed of facts,
as a house is of stones.
But a collection of facts
is no more a science
than a heap of stones is a house.*

—Henri Poincaré

The above quote illustrates a key idea that is found in all of the reform documents, and has been expressed in many different ways. In brief, a pile of facts is a curriculum that builds nowhere. We could say the same of a pile of activities or a pile of ideas. The pile of stones pictured above represents any approach that leads nowhere, that is disconnected, and makes no sense in and of itself. The question becomes: How does one build a house from a pile of stones? Or, how does one construct a house of science from a pile of facts, concepts, and theories? In short, how should students be taught?

## Building the Foundation

What do students need to construct a house of science?  First, they will need a foundation.  Of central importance in constructing a house of science is the set of abilities and understandings known as *scientific inquiry*.  The foundation of scientific inquiry, according to the *National Standards*, encompasses both understanding of scientific inquiry, and the abilities necessary to do it.  These abilities include: "asking questions, planning and conducting investigations, using appropriate tools and techniques to gather data, thinking critically and logically about relationships between evidence and explanations, constructing and analyzing alternative explanations, and communicating scientific arguments." (NSES, page 106)  These fundamental abilities and understandings are designated "Standard A," and are elaborated for each of the three grade level spans addressed by the Standards—K-4, 5-8, and 9-12. These also include science process skills, such as observing, comparing, inferring, and applying.

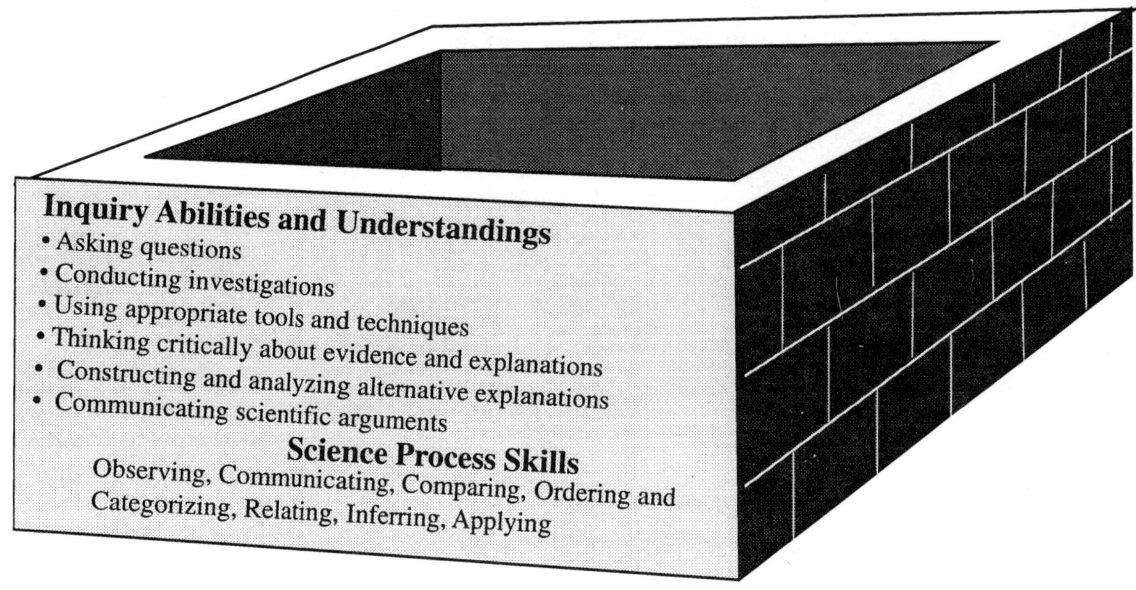


## Constructing a Framework

So, your students have a foundation; and that's an improvement over a pile of stones.  But they still need something to shape and contain those concepts and theories.  Framing is needed to help support the walls of a house.  This framing is composed of unifying principles or themes—the big ideas in science—that help students connect the individual concepts and theories.

In the house of science, these big ideas are the timbers or beams from which the framework is constructed, providing a means for supporting and organizing the pile of stones.  The framework of big ideas is intended to help students make sense of the world, which is a major goal of science.  The GEMS program has followed Project 2061 in using the term *themes* to encompass these big ideas, but one could also use the term *unifying concepts and processes* that has been used in the *National Science Education Standards* document, or *big ideas*, as used by the SS&C Project.  Definitions of commonly used themes in GEMS are included in Appendix B.

With the foundation and framework in place, your students can begin to position the individual stones to fill in the walls of the house. A student is far more likely to retain information is she has a logical place to put it—if she understands where it fits into the big picture. Within this framework, students can begin to build their own interesting and coherent structures as their houses take shape.

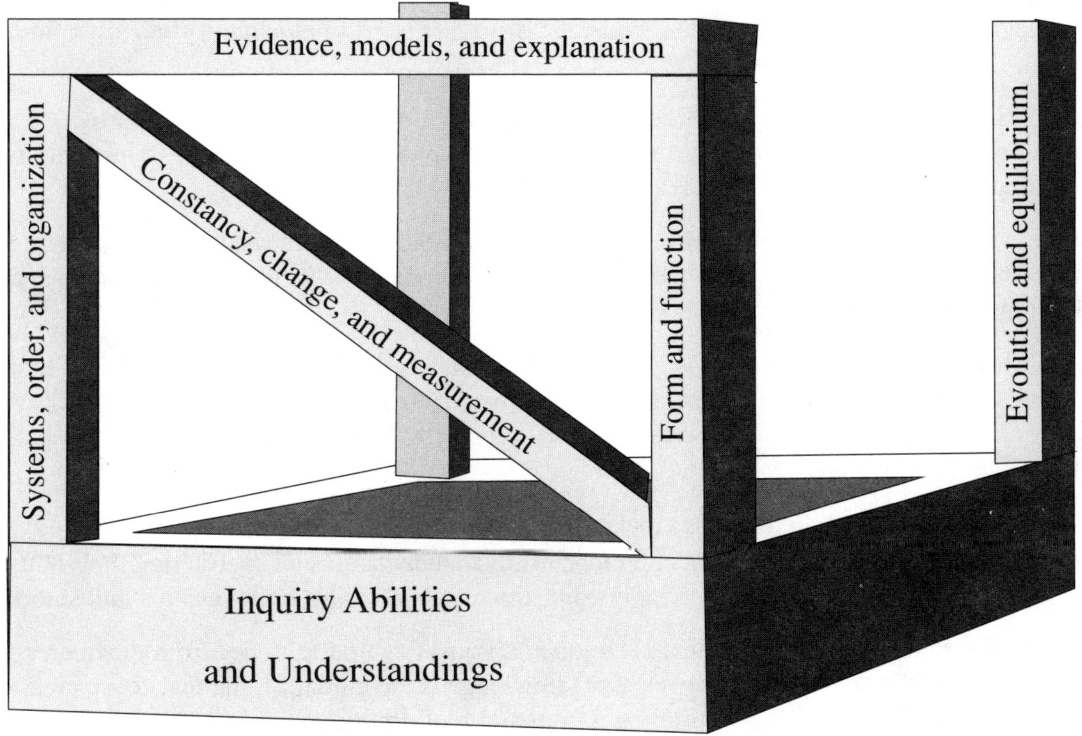



## Mortar and Nails

Of course, the stones will not stay in place by themselves. A strong wind or an earthquake could knock them out if they are not held in place with mortar and nails. In the house of science, mortar and nails represent enjoyment and curiosity. Without them, your students' houses of science won't last from year to year. Enjoyment and curiosity have kept people pursuing science for centuries. Of course young people naturally have these, but if we aren't careful they can often lose them through our educational process.

## Putting On the Roof

A vital part of the science education reform movement is the goal that *every student,* including those from groups who have been historically underrepresented in science, mathematics, and technical fields, have an opportunity to build a house of science literacy. In this metaphor the roof of the house represents the goal of achieving science literacy for all students.

# Further Insights from the "House of Science" Metaphor

**Constructivism and the Learning Cycle.** The metaphor of building a house captures an important perspective on how children learn that guides the *National Science Education Standards* and other reform documents, called "constructivism." **Based on r**esearch on child development, on the learning cycle, on the brain and how people learn, and on common misconceptions in science, the educational research community gradually converged on a model of student learning in which each individual actively constructs meaningful concepts. In this metaphor, teachers can provide students with experiences, challenges, materials and opportunities, but they cannot build houses *for* their students. **Students need to construct their own houses of science.**

**Remodeling.** It is common for learners to "remodel" their houses, as they dismantle less correct, incomplete, or out-of-date understandings with further refined and more accurate knowledge. The continual building and rebuilding that occurs is central to the constructivist approach to learning.

**Houses are never "done."** A learner's house is never done—rather it becomes a framework for lifelong learning. Scientific knowledge is expanding at a tremendous rate, and our rapidly changing world demands that we remain lifelong learners.

**All houses will look different.** The goal is not for teachers to help students construct identical-looking houses. Each learner's "house" will reflect that person's experiences, interests, and understandings. The result is the creation of a rich and diverse collection of structures that represent the diversity of human ideas, beliefs, and understandings.

**What role do scientists play?** In the house metaphor, scientists work in the stone quarry, mining, cutting, and shaping the "stones" (concepts, facts, and theories) that students use in constructing their houses. Increasingly as well, scientists have become involved in educational programs for both teachers and students.

**What role does the teacher play?** The teacher's job is not just to ship the stones from the quarry site to the classroom! Equally important as delivering science knowledge is the guidance that teachers can give students in laying a foundation for inquiry, constructing a framework of unifying concepts and processes, and ensuring that all students receive the support that they need to build their own unique houses of science. Teachers have a precious and crucial role to play as inspirers and facilitators of learning. As the famous Brazilian educator, Paulo Friere, once said, "Liberating education consists in acts of cognition, not transferrals of information."

## Whose Job Is It?

Building a house is not a simple project. It takes more than a carpenter with a few boards and nails. An entire system must be mobilized. The bank needs to provide financing. Lumber mills, quarries, and cement factories all over the country need to produce the building materials, local vendors need to bring the materials together at the building site, architects, engineers, and contractors need to obtain the training that they need to play their parts in the construction process, and building inspectors need to check on each phase of construction.

In a similar manner, a great many people have important roles to play so that science literacy can become a reality for all students. Students, teachers and administrators play the most active roles at the school site. Crucially important are parents and community members who can support their local schools by encouraging students at home, advocating for reform, and helping teachers by volunteering time, expertise, and moral support. Support from the wider system includes personnel at the State Department of Education who can provide state guidelines, workshops, and grant funding; professional societies that enable teachers to share ideas, information, and materials with others around the country; and universities who educate new teachers and provide continuing education for those already on the job. Corporate grants and local business contributions are playing an increasingly key role, recognizing the need for talented and capable future employees. Elected and career officials of government agencies are in strategic positions to encourage and support improvements in science education. In short, science education is everybody's business. In turn, positive transformation in science and mathematics education will have great benefits for everyone and for society at large.

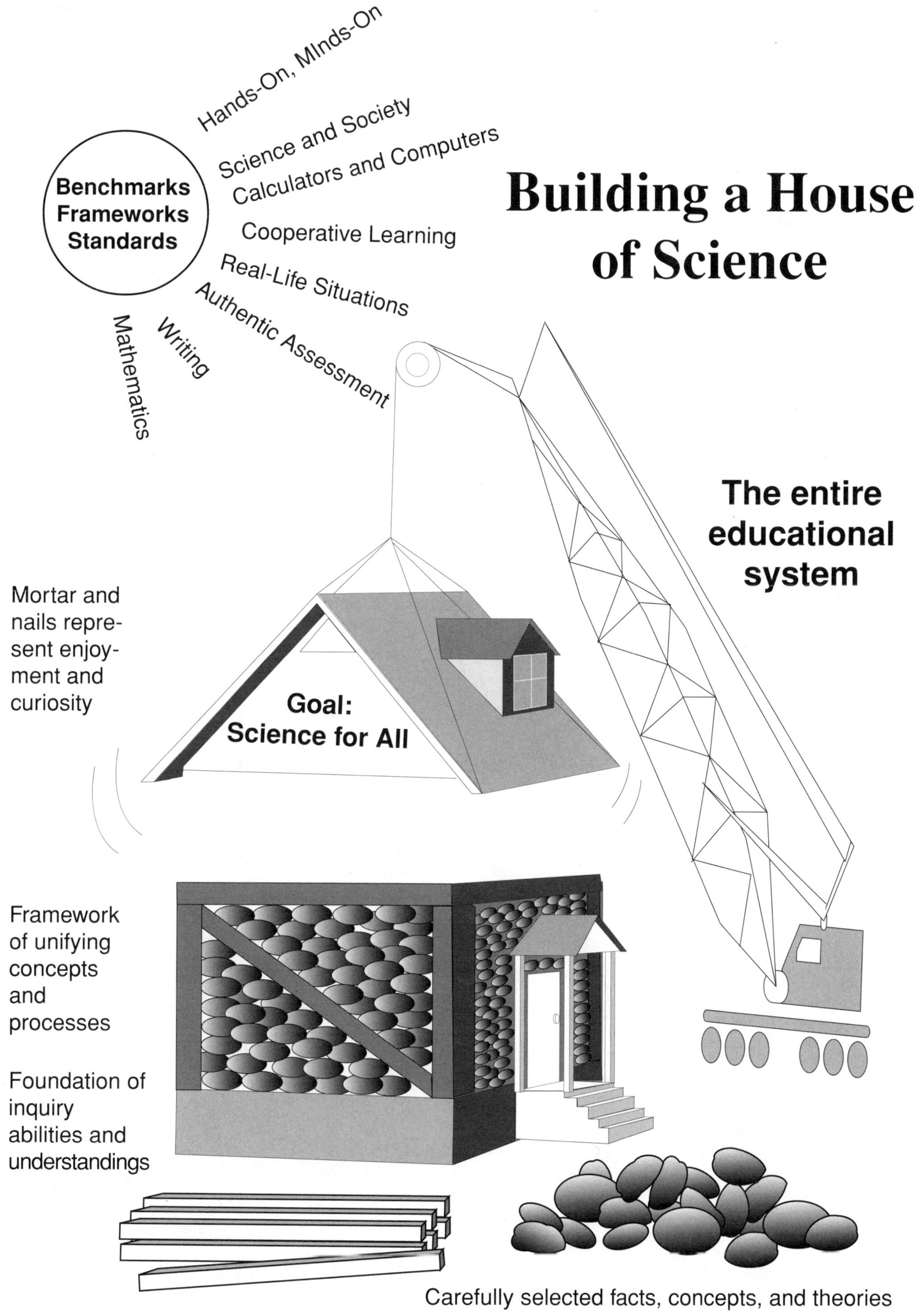

Benchmarks
Frameworks
Standards

Hands-On, Minds-On
Science and Society
Calculators and Computers
Cooperative Learning
Real-Life Situations
Authentic Assessment
Mathematics
Writing

Building a House
of Science

The entire
educational
system

Mortar and
nails repre-
sent enjoy-
ment and
curiosity

Goal:
Science for All

Framework
of unifying
concepts
and
processes

Foundation of
inquiry
abilities and
understandings

Carefully selected facts, concepts, and theories

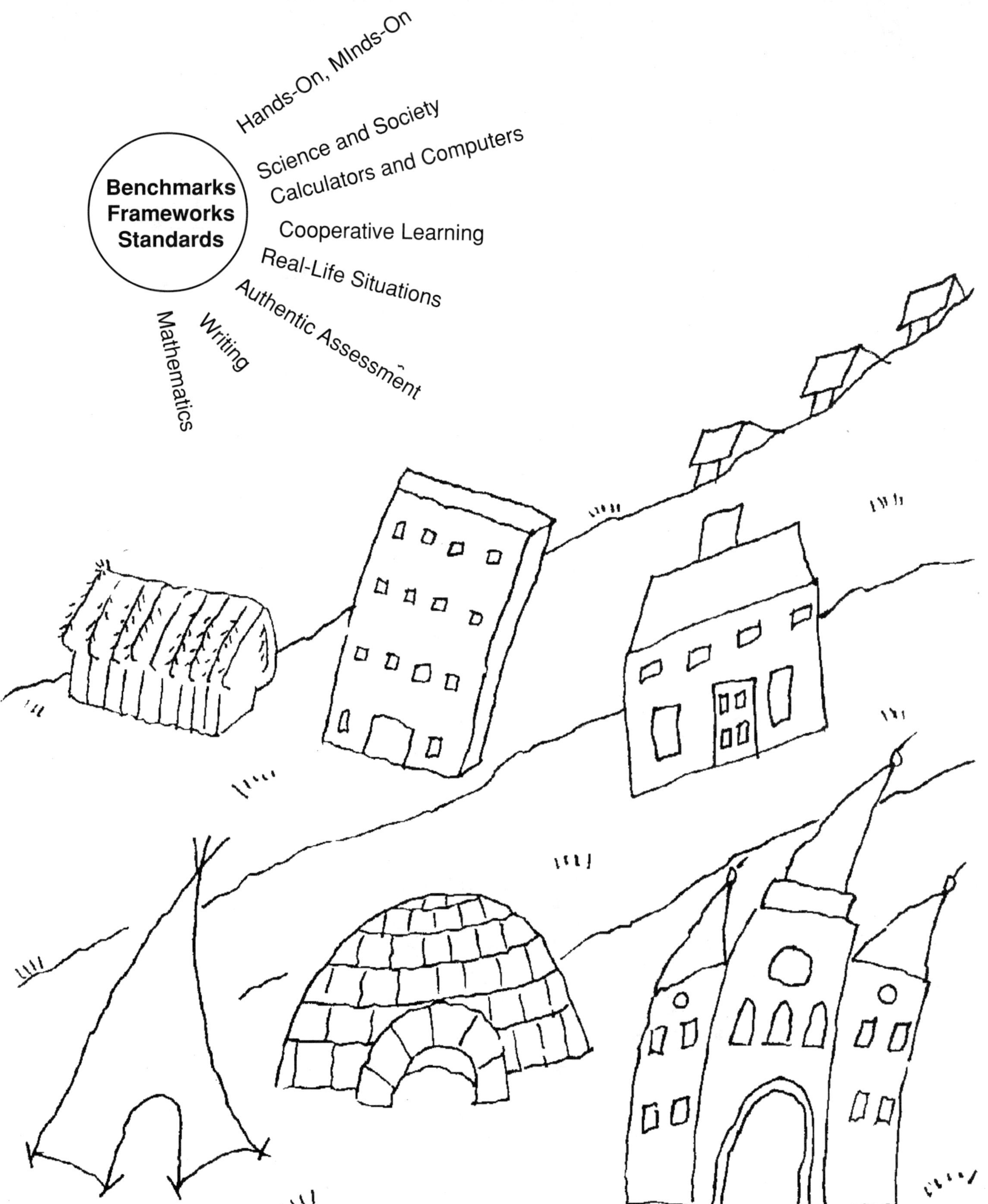


**Each child constructs her own house of science.**

# Part II

## GEMS and National Standards

This section of *The Architecture of Reform* shows how the GEMS series can be used as a strong support for the recommendations of the key reform documents. It concludes with suggestions for ways to create a plan for systemic transformation in your school or district.

**Chapter 5: GEMS and the New Reforms** describes how the GEMS series supports the recommendations of the leading reform documents.

**Chapter 6: Building a Curriculum with GEMS** discusses how GEMS units, aligned with National Standards, can be used as building blocks, either by themselves or in conjunction with other hands-on minds-on programs, to create a curriculum or provide enrichment units for each grade level, from pre-Kindergarten through the first year of high school.

**Chapter 7: Creating a Plan** provides suggestions for how to engage in local curriculum planning without having to reinvent the wheel. It also outlines systems that you will need to set up in order to implement and sustain an effective science program.

**References** This section lists all of the books and articles referred to in Parts I and II.

# Chapter 5
# GEMS and the New Reforms

One of the first and most important steps in your effort to improve the science program in your school or district is to find instructional materials that are consistent with the common vision for reform, described in Chapter 3.  The importance of this step is spelled out quite clearly in the *National Standards*:

> *Effective science curriculum materials are developed by teams of experienced teachers, scientists, and science curriculum specialists using a systematic research and development process that involves repeated cycles of design, trial teaching with children, evaluation, and revision.  Because this research and development process is labor intensive and requires considerable scientific, technical, and pedagogical expertise, teachers and school district personnel usually begin the design and development of a curriculum that meets local goals and frameworks with a careful examination of externally produced science materials.  —NSES, page 213*

Great Explorations in Math and Science (GEMS) is developed in precisely the way described, with several rounds of trial testing in classrooms all over the country, and the involvement of teachers, scientists, and curriculum developers in the careful development, revision, and production of the materials.  In fact, GEMS guides are never "finished" since teachers are always invited to write in with ideas for improvements, and revisions are made based on their comments.

While GEMS is by no means the *only* set of instructional materials that supports the new reforms, all of the GEMS guides and handbooks are consistent with the reform documents.  This is not a coincidence or a surprise.  The GEMS materials have served as models of good science education since the first Teacher's Guides were released in the mid 1980s, and several GEMS staff have served on committees to develop the new reform documents.  In this chapter, we briefly review the ways that GEMS supports the twelve aspects of a common vision of reform that were outlined in Chapter Three.

## 1. The house of science is for all students.

Providing the opportunity for all students to study science is not enough.  To engage all students it is necessary for them to be motivated to learn; and motivation is one of the great strengths of the GEMS program.  Activities that do not engage students' interest and curiosity during classroom trials are eliminated from the program.  Only activities that stimulate student interest and excitement *and* that teach pivotal science and mathematics concepts survive to become GEMS units.

Those of us who present activity-based science to students know firsthand the seemingly limitless joy that our students derive from experimenting, "messing around," and discovering.  They discover many things, including things which we did not necessarily intend for them to discover.  It is that joy of exploration that drives our students to learn.  Imagine what might be achieved if we could enable our students to retain that joy throughout their school careers—high school and college science courses would be flooded with applicants!

GEMS activities also provide easy access points so all students can succeed, increasing confidence and overcoming feelings of intimidation and avoidance that many students (and teachers) tend to associate with science.

In addition, GEMS activities invite students to be creative and to find more than one solution to a problem.  All students respond to the chance to bring their own creative energies into play.  "Going Further" activities allow students with greater interest, or teachers with more time for science, to extend the activities and take on greater challenges.


## 2. Strong commitment to diversity and equity.

Emphasis on teamwork and cooperative learning, the use of a wide variety of learning formats, and reliance on direct experience rather than textbooks makes GEMS highly appropriate for use with populations that have been historically underrepresented in science and mathematics careers.  The *Group Solutions* and *Group Solutions, Too!* GEMS guides provide highly popular and extremely effective early experience in cooperative problem solving.

In addition, a number of GEMS guides, including *Investigating Artifacts, Earth, Moon, and Stars,* and *Math Around the World*, emphasize the contributions of diverse cultures to the historical development of mathematics and science.  The guide *On Sandy Shores* incorporates proven learning modes to assist English language learners in fully participating in all the activities and many other guides take advantage of the ways that activity-based science and math lend themselves to fostering language acquisition.  The richness of multicultural contributions serves as a central criteria in the GEMS literature connections handbook.

Lawrence Hall of Science, with programs such as EQUALS and Family Math, has an impressive record of achievement in reaching historically underrepresented students and communities.  The SAVI-SELPH program, one of the antecedents of the Full Option Science System, was designed for students with physical and learning challenges.  The GEMS program grew up within this same framework of seeking to make excellent math and science programs accessible and available to all.  GEMS staff have endeavored to actively recruit and include members of groups historically underrepresented in the sciences as GEMS leaders and associates, as workshop presenters, and as teachers and students to be photographed and featured as role models in the GEMS Teacher's Guides.  The GEMS rocketry guides was designed through testing to ensure the full participation of girls/young women.  In general, the GEMS local and national testing process is carefully designed to ensure applicability and effectiveness amidst cultural and linguistic multiplicity, as well as to adjust for the wide diversity of social, economic, and geographical conditions in the United States.

# 3. Start with a foundation of inquiry abilities.

GEMS uses the "guided discovery" approach to learning. *Guided discovery* is an inclusive term. It may involve a variety of methods ranging from open-ended explorations to structured labs, simulations, role playing, and discussions. What distinguishes guided discovery from other approaches is that: 1) students have opportunities to discover for themselves, through their own experiences and reflections, something interesting about the world that they did not know before; and 2) the teacher provides guidance, helping the students progress from open-ended exploration of phenomena, to focused inquiry, experimentation, drawing of conclusions, and the construction of improved models of the world.

A classic example of open-ended inquiry is from the GEMS guide *Oobleck: What Do Scientists Do?* in which students explore a strange substance said to come from another planet. In the laboratory part of the activity students explore Oobleck in an open-ended fashion, raising their own questions and devising simple experiments, using everyday materials, to test their hypotheses. Then, the students hold a "scientific convention" during which they critically each others each others' conclusions. At this phase students discuss alternative theories—such as whether Oobleck liquefies due to heat or a release of pressure—and decide whether or not the evidence is adequate to support the team's conclusions, or if further research is necessary. Both verbal and written communication are emphasized on the scientific convention, as students revise their statements in light of new experiments and arguments. In phase three students design spacecraft to land safely on the surface of the Oobleck planet, applying what they learned about the behavior of Oobleck from the previous sessions.

Each class makes its own discoveries, based on the investigations that students choose to undertake. Teachers who present this unit to different classes over the years find that it never becomes routine. They are always amazed at the ingenuity and persistence of students for whom science comes alive.

More focused examples include GEMS guides such as *Vitamin C Testing, Hot Water and Warm Homes from Sunlight,* and *Of Cabbages and Chemistry.* In each of these units students learn a laboratory technique , and then apply the technique in a variety of situations of their own choosing.

Yet another variety of GEMS units that build students' abilities to think critically about evidence and explanations. These include the mystery series—*Crime Lab Chemistry, Fingerprinting,* and *Mystery Festival.*

As specified in the *National Science Education Standards*, GEMS activities are minds-on as well as hands-on. And, they provide opportunities for students to integrate process skills with pivotal science concepts. Both the inquiry abilities and science concepts are listed on the title page of each GEMS Teacher's Guide.

---

**Inquiry Abilities and Understandings** (from the *National Science Education Standards*)
• Asking questions
• Conducting investigations
• Using appropriate tools and techniques
• Thinking critically about evidence and explanations
• Constructing and analyzing alternative explanations
• Communicating scientific arguments

---

# 4. Construct a framework of unifying concepts and processes

*Science for All Americans* introduced the idea of "themes" as broad, unifying principles that connect the different science disciplines. The *SS&C Content Core* referred to "great ideas" which underlie and connect the sciences. And the *National Standards* devoted the first part of its chapter on content to "unifying concepts and processes." All of these terms refer to similar broad, overarching ideas that link various disciplines. Themes are examples of logical ways to organize science. As such, they can be used to create the frame that helps students structure the learning of new concepts and theories—supporting and giving shape to the "pile of stones" discussed in Chapter 4.

Consider, for example, the value of learning about Patterns of Change. In all disciplines, scientists look for patterns in the data that they have collected. They may look for an increase over time, or a decrease, or a regular cyclic pattern, or a non-repeating random pattern. Any and all of these patterns of change provide important information about the world they are studying. The more students know about the patterns they might find, the more they are likely to learn from new situations. Cyclic patterns, for instance, are especially revealing. Whether you're in first grade learning about the water cycle, or the seasons, or the life cycle of a butterfly—those are all cycles. Or later, when you learn about the phases of the moon—ah, another cycle! Even later in high school or college when you learn about something in human physiology called the Krebs cycle, you'll have a good start in understanding it because you already know what a cycle is. Cycles will appear again and again in all the sciences, on very short and very long time scales, involving matter or energy or both. Each time you encounter a new kind of cycle you will deepen your understanding of cyclic change. And the more you understand about cyclic change, the more you will be able to learn about the world.

In using the word "themes," there is some danger of confusion with the idea of organizing the curriculum around a particular subject, such as whales, or space travel. While selecting a subject matter theme can be a useful way to relate science to other subjects, it is not what is meant by "themes" in science education reform.

Above, we've mentioned Patterns of Change as one theme that unites the sciences. There are certainly others; but there is no single, perfect list of themes; and each of the reform documents describes them a little differently. We have selected ten of the most commonly used themes to provide a framework for the GEMS series. These are described in Appendix B, and are listed in the margin on this page.

## GEMS Themes

- Patterns of Change
- Stability
- Energy
- Matter
- Scale
- Structure
- Systems and Interactions
- Diversity and Unity
- Evolution
- Models and Simulations

## Unifying Concepts & Processes

**from *National Science Education Standards, pages 115-119***

- Systems, Order, and Organization
- Evidence, models, and explanation
- Constancy, change, and measurement
- Evolution and equilibrium
- Form and function

## 5. Identify the most important concepts, theories, and skills.

As mentioned earlier in this handbook, most of the topics that have been included in the *Benchmarks for Science Literacy* and the *National Science Education Standards* are very similar, although they are arranged somewhat differently in the two documents.  That's good news for school district planners.  It means that there is widespread agreement among scientists and educators about what students should know and be able to do.  You will find that a great many of these fundamental concepts and theories are presented, through a guided discovery approach in the GEMS Teacher's Guides.

There is also widespread agreement about the process skills that students must learn through hands-on experiences.  These skills, which form the foundation of the House of Science discussed in the last chapter, include observation, recording, experimenting, and so on.

Identifying which key concepts, theories, and process skills are emphasized in each GEMS guide is easy.  Just look at the first page in each GEMS guide and you'll find lists of the most important ideas and skills emphasized in the unit, under the following categories:

• Inquiry Abilities and Process Skills

• Science concepts and theories

• Themes

• Nature of Science

• Mathematics Strands

The key ideas listed within each of these categories can be found in just about all of the reform documents in one form or another.  By planning a sequence of GEMS units, as illustrated in Chapter 7, it is possible to help your students learn a significant number of the most important concepts, theories, and process skills recommended for their grade level, without sacrificing the inquiry approach.

# 6. Integrate ideas from different fields whenever possible.

Another key idea found in virtually all of the science education reform documents is that science should not stand alone, isolated from other subjects; but should be integrated with the rest of the curriculum and the real world of fascinating phenomena and human concerns.

A great many of the GEMS guides integrate knowledge and skills from different disciplines.  For example, in *Bubble-ology*, students conduct controlled experiments designed to determine which bubble solution makes the biggest bubble.  They measure each trial, then calculate the mean, median, and range of each set of data.  Finally, they interpret the data, drawing conclusions based on their mathematical analyses.  This example illustrates that science and mathematics can be seamlessly united into a single lesson.  In addition to its mathematical component, Bubble-ology integrates physics, chemistry, an aerodynamics.  There are many such examples.  Many GEMS guides integrate mathematics and science.  Here are a number of GEMS guides which integrate two or more disciplines:

> *Color Analyzers* integrates the physics of light and color and the biology of color perception.

> *Vitamin C Testing* integrates chemistry, nutrition, and consumer science.

> *Convection: A Current Event* demonstrates the value of physics in understanding weather and oceanography.

> *Moons of Jupiter* combines astronomy and space science with the history of science and the mathematics of scale.

> *Frog Math* is a synthesis of children's literature and mathematics.

> *Global Warming and the Greenhouse Effect* and *Acid Rain* combine atmospheric chemistry and social studies with environmental studies and many other mathematical and biological aspects.

> *Investigating Artifacts* blends mathematics, astronomy, literature, archaeology, and cultural anthropology, with strong emphasis on Native American and world cultures.

> *Learning About Learning* involves students in using science to investigate how they learn, combining current research about the human brain with articles on the history of scientific and medical research, biology, psychology, and important issues of health and safety.

The *GEMS Literature Handbook* is an especially powerful tool designed to help teachers relate science to language, social studies, and other subject areas.  It includes a wide variety of ideas for how to relate science and literature, and provides an annotated literature connections list for each of the GEMS Teacher's Guides.

## 7. Illustrate the history and nature of science.

GEMS Guides suggest a myriad of ways to involve students in learning about the nature of science by doing it. In many cases, they reflect on what they've done to see how it is like the work of professional scientists. For example:

> In *Investigating Artifacts,* the students learn about the work of archaeologists and anthropologists by excavating middens and creating artifacts.

> In *Mapping Animal Movements*, they learn techniques used by field biologists, and then compare their work with an actual study in which a similar technique was used to study deer in the wild.

> In *Fingerprinting, Crime Lab Chemistry,* and *Mystery Festival,* the students compare the results of their own investigations with the work of actual forensic scientists.

> *In Learning About Learning*, the students investigate their own learning processes, and compare their discoveries with the work of actual psychologists and brain researchers. They also learn about the history of government regulation of harmful chemicals and the role played by scientific researchers in occupational safety.

In each case, the connection between the students' activities and the work of professional scientists is made as explicit as possible.

Other guides introduce the history of science as a way of helping students understand how science has evolved over time. For example, in *Earth, Moon, and Stars*, students start by inventing their own flat-earth models to explain the phenomena of night and day. In the second activity they explore the ancient Greek idea that the Earth is shaped like a sphere. Throughout the unit, students apply reasoning and critical thinking to the gradual emergence of modern ideas about the Earth's shape and gravity. *Moons of Jupiter* opens with a student-centered re-creation of Galileo's famous observations of four of Jupiter's moons, bringing a famous episode in the history of science to life!

Since the nature of science is considered such an important aspect of the GEMS series, we have listed the major qualities of science emphasized in a given guide on the title page, along with concepts, inquiry skills, and themes. A description of the various aspects of science portrayed in GEMS guides is included in Appendix C of this handbook.

## 8. Show how science and technology inform societal issues.

As stated in the *National Science Education Standards* (page 107), "An important purpose of science education is to give students a means to understand and act on personal and social issues." It is important, therefore, to help students relate science to social issues  while they are in school so that later, when they become voting citizens, they will be able to make the connections on their own.

Two of the Teacher's Guides in the GEMS series are full-fledged units on societal issues that are significantly related to science and technology: *Acid Rain,* and *Global Warming and the Greenhouse Effect.* Both of these units begin with what students know or have heard about these topics, and progress to activities that help students understand the scientific theories and findings that underlie these issues, as well as the questions of social policy that are faced by decision-makers and citizens. Neither guide takes a position on social policy, but students are encouraged to recognize what the issues are, to debate the issues from different points of view, and to formulate their own opinions.

Different kinds of activities are used in these units to help students understand the social issues from different perspectives. In *Acid Rain*, the students take on different roles in a skit about the effect that acid rain has had on a lake in which they live. In *Global Warming and the Greenhouse Effect* the students read a story about a boy growing up on a coral atoll, where the impact of global warming—should it occur—will be much greater than on people who live in the interior of a continent.

Other GEMS guides that involve students in applying science to societal issues include:

> *Hot Water and Warm Homes from Sunlight,* in which students learn how solar energy can help reduce our dependence on fossil fuels.

> *River Cutters* enables students to observe how ground water pollution and toxic waste can spread through river systems. This unit also explores the impact of dam construction on river systems.

> *On Sandy Shores* invites students to investigate the effects that human habitation has had on marine ecosystems.

> *Sifting Through Science*  includes a culminating hands-on activity on separating "garbage" that introduces young students to some basic ideas about recycling and reuse.

> *Investigating Artifacts*, when presented to older students, includes controversial issues regarding what happens when Native American artifacts are discovered, for example, at a planned construction site for a new shopping mall. Several of the activities in the guide also explore issues of cultural relativism and respect for diverse cultures.

> *Learning About Learning*  provides multiple experiences for students gather information that allows them to have an open-ended discussion and analyze issues related to animal and human experimentation. The guide also introduces the important social and educational concept of multiple intelligences.

## 9. Each student needs to construct his or her own house of science.

GEMS is based on a constructivist theory of learning.  Many units begin by asking the students open-ended questions to determine their current thinking on the topic they are about to study.  In the GEMS unit *Acid Rain*, this becomes an interesting activity in itself, called "Pick Your Brain About Acid Rain."

As the activities in *Acid Rain* progress, students are encouraged to interpret data from experiments and to more clearly articulate how the results support or refute their personal models and explanations of natural phenomena.  They are also encouraged to listen to the ideas of others, looking for points of agreement and disagreement.  At all times, students are encouraged to make up their own minds, even if it means disagreeing with the majority of the class or even with the teacher!

In other GEMS units, students think about the rules of evidence and inference in their arguments and debates.  For example, in the forensic science series, *Fingerprinting*, *Crime Lab Chemistry,* and *Mystery Festival,* the students reflect on how the rules of evidence and inference are different in deciding whether or not someone should be convicted of a crime, than in deciding on the merits of a scientific theory.

In *Learning About Learning*, students work together on a solving a maze to learn more about their own learning strengths.  As they proceed through the unit they learn more about different learning styles and multiple intelligences, and how learning is related to survival, health, and safety.  They also learn how brain research shows that stimulation leads to brain cell growth and consider the famous case of Genie, who suffered serious abuse and deprivation as a child.  The learning focus and activities give students a strong consciousness of how they are constructing their own houses of science!

What is most important in each of these examples is that the students are given opportunities to consider alternative ideas and come to their own conclusions, so that learning is deep and long-lasting.

## 10. Students need to collaborate in teams much of the time.

In *all* GEMS units students are encouraged to work together to explore a phenomenon, or solve a mystery, rather than engage in negatively competitive behavior.  For example, in *Group Solutions,* student teams solve a variety of puzzles and problems in which each member of the team has a part of the solution.  If the team does not cooperate, they cannot solve the problem.

Emphasis on teamwork and cooperative learning helps to form a bridge between students with different educational backgrounds, cultures, and interests, and helps students learn from other students.  Team work is especially valuable in improving language and communication skills.  Group work during the investigations not only helps students learn science; it also helps to prepare them for workplaces of the future.

Suggestions are provided in each GEMS Teacher's Guide for how best to organize the class for each activity, depending on the materials that are needed, the procedures that are to be performed, and the objectives of each activity. Suggestions are also provided for the pooling of data from several groups, and questions for guiding class discussion.

## 11. Align assessment with curriculum and teaching.

The *National Science Education Standards* was conceived with the idea of setting standards for science content, teaching, *and* assessment, because it was recognized that assessment was absolutely essential to determine whether or not the new reforms were implemented and working.

The GEMS program has created a handbook on assessment of hands-on, minds-on activities.  Entitled *Insights and Outcomes: Assessments for Great Explorations in Math and Science* (1995), the handbook provides teachers and administrators with tools to create a science assessment program for their schools and districts.  *Insights and Outcomes* provided a great many concrete examples of strategies for assessment, including authentic test items and ideas for portfolio exhibits.  The handbook shows how assessment tasks are directly related to expected outcomes, and the kinds of evidence that teachers can look for to see if they are meeting their goals.

For example, one of the expected outcomes of the GEMS unit *Experimenting With Model Rockets* is that students learn to control variables.  To find out what students learn from this unit, assessment tools are provided in which students are asked to critique, plan, and conduct controlled experiments on entirely different topics.  A wide variety of modes, from writing letters and critically examining newspaper articles, to investigating real materials, are used to find out how students can apply what they are learning in the GEMS units.

## 12. Mobilize the whole system to support the Reforms

The creators of GEMS recognize that in the era of declining school budgets it is not easy to find funds for professional development and for science supplies and equipment. Consequently, GEMS guides are developed so that a minimum of expense is necessary to support good science instruction. That means inexpensive, easy-to-obtain materials, and simple, clear instructions, so that teachers can do the activities even if a workshop is not available.

Nonetheless, good instructional materials and enthusiasm are not enough to change science education. Support is definitely needed from administrators and the community. Even inexpensive materials require a budget. Preparing for hands-on science classes takes time, and coordinating the school curriculum takes even more time. Teachers need to be given time to do these things as part of their school day. Additionally, teachers need to feel that they are part of a larger network of professionals, and that means opportunities to communicate at conferences and via computer networks. Adequate pay and classes that are not too large are also important aspects of reform. In other words, the *system* needs to support improvement in science education.

The GEMS approach makes it easier to acquire the financial support required for start-up, simply because fewer resources are needed to achieve quality science education, than are needed to start many other programs.

## Conclusion

In summary, GEMS Teacher's Guides are consistent with the broad vision of the new science education reform movement that is sweeping the country with support from leading science organizations, and state and federal governments. The GEMS Guides are therefore valuable tools, but they alone are of course not sufficient. Along with transformations in national and state educational establishments and resources, local support and planning is also needed.

In the next chapter we describe how sequences of GEMS units can be used either as essential parts of a core curriculum, or as enhancement units to supplement other inquiry-based instructional materials. We relate these sequences to the *National Science Education Standards* to demonstrate more specifically that GEMS units and sequences are excellent resources for implementing the recommendations of the *National Standards*.

# Chapter 6

# Building a Curriculum with GEMS

As described in the previous chapter, all of the GEMS Teacher's Guides support the common vision of science education reform. While each of the units can stand alone, students will learn much more from thoughtfully planned sequences and combinations of units. This is the idea of *synergy*—that the whole is greater than the sum of the parts.

Sequences of two, three, or four GEMS units can become building blocks in a core curriculum plan that is designed to meet the requirements of a particular school or school district. The term *core curriculum* usually refers to units that are to be taught by all teachers; while *enrichment units* are usually considered optional. GEMS units and sequences can be used either as core or enrichment materials in a program that is based on a national, state, or local guidelines.

Just as building blocks can be used in a variety of ways to construct buildings that differ in shape or texture, GEMS sequences can also be used in a variety of ways. A sequence can be used to emphasize a particular set of content standards during a two or three month period. Or, a sequence of units can be spread over a two or three year period, so that understanding can be built gradually, over time. As an example, we'll consider a sequence of GEMS units that brings together standards on "Scientific Inquiry" and "Earth and Space Science."

To envision how this sequence might be used, look over the summary on the next page, which shows how each unit reflects the *National Science Education Standards*. Then imagine how *you* would use it to build your curriculum.

# *An Astronomy & Space Science Sequence for Grades 5-8*

A unit on the subject of astronomy at the upper elementary or middle school level might consist of three GEMS guides: *Oobleck: What Do Scientists Do? Earth, Moon, and Stars,* and *Moons of Jupiter.* For a class that met every day, these units could be presented over a period of one to two months.

**1)** *Oobleck: What Do Scientists Do?* Grades 4-8. Students investigate and analyze the properties of a strange green substance said to come from another planet. The class holds a scientific convention to critically discuss their scientific findings; and design a spacecraft to land on an ocean of Oobleck. In the final session, they compare the methods they used in the laboratory, discussion, and spacecraft design phase, to the work of professional scientists, such as those on the Mars Viking mission.

**2)** *Earth, Moon, and Stars,* Grades 5-8. In this extensive unit, students learn a great deal about the Earth's shape and gravity, moon phases and eclipses, and what the apparent movements of the stars tell us about the Earth. Students observe the moon and stars in the sky, create models to explain their observations, and critically discuss their ideas.

**3)** *Moons of Jupiter,* Grades 4-9. In this unit your students become "Galileos" as they re-create, through viewing beautiful slides (provided with the guide) Galileo's historic telescope observations of Jupiter's moons. They observe and record orbits of the moons over time and learn why these observations helped signal the birth of modern astronomy. Students experiment to see how craters are formed, construct scale models to better understand the vast sizes and distances in the Jupiter system, and view Jupiter's major moons as imaged by modern space probes. In the final session, teams work together to create a model of a settlement for explorers to one of Jupiter's moons.

These three units can be presented sequentially at one grade level, or they can be presented in subsequent years, so that learning builds from year to year. These two approaches are diagrammed below:

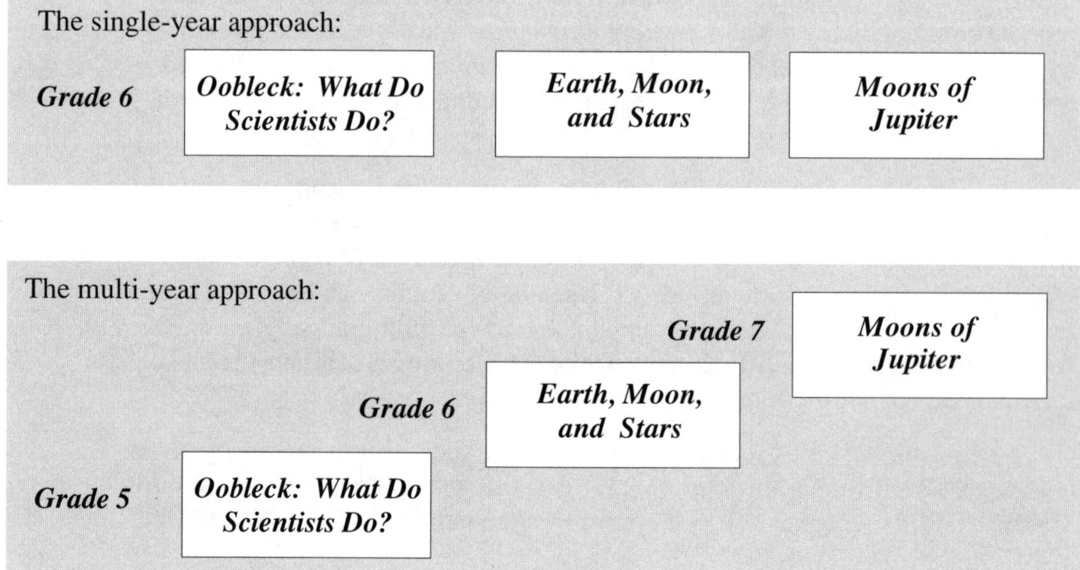

# *National Standards Addressed By This Sequence*

Following is a brief overview of how this sequence of GEMS units addresses the content standards for grades 5-8 in Chapter 6 of the *National Science Education Standards* (pages 143-170). National Standards are shown in italics; discussion of student activities in this sequence are shown in regular type.

***Science as Inquiry****: As a result of activities in Grades 5-8, all students should develop:*
- *Abilities necessary to do scientific inquiry*
- *Understandings about scientific inquiry*

In *Oobleck: What Do Scientists Do?* Students start by investigating this strange substance to find out its properties. In doing so, they raise interesting questions. For example: "Does Oobleck melt from the heat of my hand, or does it turn to a liquid when I release the pressure?" They use a variety of tools to conduct experiments in order to answer their own questions, describe their results in writing, critically examine other students' findings, and consider alternative explanations for the odd behavior of Oobleck. In *Earth, Moon, and Stars* the students use observations of the sky and modeling to investigate lunar phases and eclipses; and In *Moons of Jupiter* they conduct investigations to explain some of the impact features that they observe on the moons of Earth and Jupiter.

***Earth and Space Science.****  As a result of their activities in Grades 5-8, all students should develop an understanding of... Earth in the solar system:*

In *Earth, Moon, and Stars*, students critically debate alternative theories about the Earth's shape and gravity. They also gather data on the moon's position and changing shape over a month, and create a model to explain the daily movement of the sun, moon, and stars, and both phases and eclipses of the moon and solar eclipses. In *Moons of Jupiter*, students observe moons orbiting Jupiter, and interpret the data to find the periods of the moons. They also model both the Jupiter and Earth-Moon system using the same scale for size and distance.

***Science and Technology.****  As a result of activities in Grades 5-8, all students should develop :*
- *Abilities of technological design*
- *Understandings abut science and technology*

In *Oobleck: What Do Scientists Do?* Students design a spacecraft to land on an ocean of Oobleck. They also discuss how they behaved as both scientists (in the laboratory and discussion activities) and as engineers (in the spacecraft design activity). In *Moons of Jupiter*, students design settlements where explorers live while they explore one of the Jupiter's moons. In both units students solve creative design problems and discuss the ways that science and technology complement each other.

***History and Nature of Science.****  As a result of activities in grades 5-8, all students should develop understanding of*
- *Science as a human endeavor*
  - *Nature of science*
  - *History of science*

In *Earth, Moon, and Stars*, students learn about some of the earliest theories of the Earth and its place in the universe, and how these ancient myths formed the roots of modern science. In *Moons of Jupiter*, students replicate Galileo's observations of Jupiter's moons, and see how he figured out their periods. In *Oobleck: What Do Scientists Do?* they hold a scientific convention and compare their own investigations with the recent history of the Viking project.

***Unifying Concepts and Principles.****  Activities in all three guides emphasize:*
- *Constancy, change and measurement (e.g. observations of the moon's phases)*
- *Evidence, models, and explanation (e.g. modeling moon phases and eclipses)*
- *Scale and Structure (e.g. the vast scale of the Jupiter system in contrast to the Earth-Moon system.)*

# More GEMS Sequences

Just as flexibility is a quality of individual GEMS units, it is also a quality of GEMS sequences. Many different combinations of guides are possible, to meet a variety of different Standards and educational goals. And any given sequence can be used as part of a core curriculum plan, or for enrichment. On these two pages you will find a list of sequences that have been devised by the GEMS author team. Each sequence is recommended for a given grade level, but in most cases the sequence could be used a year or two above or below the suggested level. Each sequence is described in greater detail in the remainder of this chapter.

**Preschool**

**Life Science Sequence**

*1) Tree Homes; 2) Eggs Eggs Everywhere; 3) Penguins and Their Young*

**Kindergarten**

**Life Science Sequence I**

*1) Ladybugs; 2) Ant Homes Under the Ground; 3) Buzzing a Hive*

**Life Science Sequence II**

*1) Hide a Butterfly; 2) Animal Defenses*

**Grade 1**

**Physical Science Sequence**

*1) Sifting Through Science; 2) Liquid Explorations; 3) Bubble Festival; 4) Involving Dissolving; 5) Secret Formulas*

**Grade 2**

**Integrating Math, Science, Literature, and Social Studies**

*1) Treasure Boxes; 2) Frog Math; 3) Investigating Artifacts*

**Grade 3**

**Chemistry Sequence**

*1) Secret Formulas; 2) Mystery Festival*

**Grade 4**

**Ecology Sequence**

*1) On Sandy Shores; 2) Terrarium Habitats; 3) Schoolyard Ecology*

**Evidence and Inference Sequence**

*1) Crime Lab Chemistry; 2) Fingerprinting; 3) Mystery Festival; 4) Investigating Artifacts*

| **Grade 5** | **Physical Science Sequence** |
| --- | --- |
| | *1) Bubble-ology; 2) Bubble Festival* |
| | **Controlled Experiments Sequence** |
| | *1) Hot Water and Warm Homes from Sunlight;* |
| | *2) Paper Towel Testing; 3) Vitamin C Testing* |

| **Grade 6** | **Earth Science Sequence** |
| --- | --- |
| | *1) Stories in Stone; 2) River Cutters* |
| | **Chemistry Sequence** |
| | *1) Chemical Reactions; 2) Vitamin C Testing;* |
| | *3) Of Cabbages and Chemistry; 4) Acid Rain* |
| | **Math, Science, and Technology  Sequence** |
| | *1) Height-O-Meters; 2) Experimenting With Model Rockets* |

| **Grade 7** | **Astronomy and Space Science Sequence** |
| --- | --- |
| | *1) Oobleck: What Do Scientists Do? 2) Earth, Moon, and Stars;* |
| | *3) Moons of Jupiter* |
| | **Scientific Instruments Sequence** |
| | *1) More Than Magnifiers; 2) Color Analyzers;* |
| | *3) Microscopic Explorations* |
| | **Ecology Sequence** |
| | *1) Acid Rain; 2) Global Warming* |

| **Grade 8** | **Life Science Sequence** |
| --- | --- |
| | *1) Animals in Action; 2) Mapping Animal Movements;* |
| | *3) Mapping Fish Habitats; 4) Learning About Learning* |
| | **Physical Science Sequence** |
| | *1) Discovering Density; 2) Convection: A Current Event* |

# Brief Summaries of GEMS Sequences

This section briefly summarizes the sequences listed on the previous two pages. Included in each summary is a summary of the main *National Science Education Standards* addressed by each sequence.

---

**Preschool Life Science Sequence**

Learning About Animals and Where They Live

**1)** *Tree Homes*, Preschool-1—The activities in this guide encourage appreciation for trees and animals that live in tree homes, stimulating children's interest in the world around them and the biological need for warmth and shelter.

**2)** *Eggs Eggs Everywhere*, Preschool-1—This guide introduces children to the wonder of eggs and provides them with experience to begin developing key concepts in biology and life sciences.

**3)** *Penguins and Their Young*, Preschool-1—Children learn about the emperor penguin, including its body structure, its cold home of ice and water, what it eats, and how emperor penguin parents care for their young.

**National Standards Addressed by this Sequence:**
- Life Science Standard, K-4: The characteristics of organisms; Life cycles of organisms; Organisms and their environment
- Inquiry Standard, K-4: Abilities necessary to do scientific inquiry
- Science in Personal and Social Perspectives, K-4: Changes in environments.
- Unifying Concepts and Principles: Form and Function

---

**Kindergarten Life Science Sequence I**

Studying Insect Life and Life Cycles

**1)** *Ladybugs*, Preschool-1—This unit takes off from childhood delight with ladybugs to present key science and math concepts to young children, including basic information relating to animal life and ecology.

**2)** *Ant Homes Under the Ground*, Preschool-1—This set of activities delves into many aspects of these ubiquitous and fascinating social insects. Students learn about ant body structure, jobs, and homes by observing ants in nature and in an ant farm.

**3)** *Buzzing a Hive,* Kindergarten-3—In this extensive and fascinating unit, students learn about the complex social behavior, communication, and hive environment of the honeybee.

**National Standards Addressed by this Sequence:**
- Life Science Standard, K-4: The characteristics of organisms; Life cycles of organisms; Organisms and their environment
- Inquiry Standard, K-4: Abilities necessary to do scientific inquiry
- Science in Personal and Social Perspectives, K-4: changes in environments
- Unifying Concepts and Principles: Systems, Order, and Organization

<table>
<tr><td valign="top" width="28%">

**Kindergarten Life Science Sequence II**

Simulating Animal Adaptation

</td><td valign="top">

**1)** ***Hide a Butterfly***, Preschool-K—Children create camouflaged butterflies, hungry birds, and a meadow of flowers to enact "The Butterfly Play," and learn the basic concepts of protective coloration.

**2)** ***Animal Defenses*** Preschool-K—In these activities, children add defensive structures to an imaginary defenseless animal.  Then, in classroom dramas, their defended animals encounter a Tyrannosaurus rex.  In other activities they extend the concept of defenses from structures to behaviors in animals and people.

**National Standards Addressed by this Sequence:**
- Life Science Standard, K-4:  The characteristics of organisms; Organisms and their environment
- Inquiry Standard, K-4: Abilities necessary to do scientific inquiry
- Unifying Concepts and Principles: Evolution

</td></tr>
<tr><td valign="top">

**Grade 1 Physical Science Sequence**

Investigating the Properties of Materials

</td><td valign="top">

**1)** ***Sifting Through Science***, Kindergarten-2—This elegant physical science unit constructs learning and encourages its application.  Three free-exploration learning stations build to a concluding whole-class activity in which students apply what they've learned about properties of materials.

**2)** ***Liquid Explorations,*** Grades 1-3—In this series of fun and engaging activities, students explore the properties of liquids.  They play a classification game, observe how food coloring moves through different liquids, then create secret salad dressing recipes and an "Ocean in a Bottle."

**3)** ***Bubble Festival,*** Kindergarten-6—This guide includes 12 classroom table-top activities with set-up instructions.  Learning stations encourage independent thinking and cooperative learning—bring fun and excitement to the classroom, while "bubbling over" with science and math.

**4)** ***Involving Dissolving,*** Grades 1-3—In this series of highly "involving" activities, students learn about the concepts of dissolving, evaporation, and crystallization.

**5)** ***Secret Formulas,*** Grades 1-3—In these wildly popular, sweet, and tasty activities, students investigate the properties of substances as they make their own personal brands of paste, toothpaste, cola, and ice cream.

**National Standards Addressed by this Sequence:**
- Physical Science Standard, K-4:  The properties of objects and materials
- Inquiry Standard, K-4: Abilities necessary to do scientific inquiry
- Science and Technology Standard, K-4: Abilities of technological design
- Unifying Concepts and Principles: Constancy, change, and measurement

</td></tr>
</table>

<table>
<tr><td valign="top" width="30%">

**Grade 2
Integrating
Math, Science,
Literature, and
Social Studies**

Learning from the
Objects Around Us

</td><td valign="top">

**1)** *Treasure Boxes*, Kindergarten-3—This compelling set of activities connects mathematics and literature with the spellbinding interest engendered by collections of small everyday objects (the "treasures").

**2)** *Frog Math,* Kindergarten-3—In an artful interweaving of mathematics and literature, this series of lively mathematics activities jumps off from one of the well-known "Frog and Toad" stories, "The Lost Button."

**3)** *Investigating Artifacts: Making Masks, Creating Myths, Exploring Middens,* Kindergarten-6—This guide weaves together three activities related to anthropology and archaeology and to diverse Native American and world cultures.

**National Standards Addressed by this Sequence:**
- History and Nature of Science Standard, K-4: Science as a human endeavor
- Inquiry Standard, K-4: Abilities necessary to do scientific inquiry
- Unifying Concepts and Principles: Systems, order and organization

</td></tr>
</table>

<table>
<tr><td valign="top" width="30%">

**Grade 3
Chemistry
Sequence**

Using Chemistry
for Problem
Solving and
Creativity

</td><td valign="top">

**1)** *Secret Formulas,* Grades 1-3—In these wildly popular, sweet, and tasty activities, students investigate the properties of substances as they make their own personal brands of paste, toothpaste, cola, and ice cream.

**2)** *Mystery Festival,* Grades 2-8—This exciting guide features two imaginative and compelling mysteries, one for younger and one for older students. Students observe the "crime scene" then conduct crime lab tests on the evidence and analyze the results to solve the mystery.

**National Standards Addressed by this Sequence:**
- Science in Personal and Social Perspectives Standard K-4:  Changes in environments
- Physical Science Standard, K-4: Properties of objects and materials
- Science and Technology Standard, K-4: Abilities of technological design
- History and Nature of Science Standard, K-4: Science as a human endeavor
- Inquiry Standard, K-4: Abilities necessary to do scientific inquiry
- Unifying Concepts and Principles: Evidence, models, and explanation

</td></tr>
</table>

<table>
<tr><td valign="top">

**Grade 4<br>Ecology<br>Sequence**

Studying the
Interactions of
Plants, Animals, and
the Environment

</td><td valign="top">

**1)** *On Sandy Shores*, Grades 2-4—Students explore and deepen their understanding of many aspects of the "sandy shore," from grains of sand and diverse animals to more complex biological and ecological interactions.

**2)** *Terrarium Habitats,* Kindergarten-6—These activities bring the natural world into your classroom and deepen student understanding of and connection to all living things.

**3)** *Schoolyard Ecology,* Grades 3-6—The schoolyard, be it rural green or urban asphalt, is the environment to be investigated.  Many ecological, environmental, and life science concepts are explored.

**National Standards Addressed by this Sequence:**
- Science in Personal and Social Perspectives Standard K-4:   Changes in environments
- Life Science Standard, K-4: Organisms and their environments
- Earth and Space Science :Standard, K-4: Properties of earth materials
- Inquiry Standard, K-4: Abilities necessary to do scientific inquiry
- Unifying Concepts and Principles: Systems, order and organization; Evolution and equilibrium; Evidence, models, and explanation

</td></tr>
<tr><td valign="top">

**Grade 4<br>Evidence and<br>Inference<br>Sequence**

Using Data to
Construct a
Reasonable
Explanation

</td><td valign="top">

**1)** *Crime Lab Chemistry*, Grades 4-8—Challenged to determine which of several black pens was used to write a ransom note, students learn and use paper chromatography, as they explore the concepts of solubility, pigments, and separation of mixtures.

**2)** *Fingerprinting,* Grades 4-8—Students explore the similarities and variations of fingerprints in these "fingers-on" activities.  Students take their own fingerprints, devise their own classification categories, then apply their classification skills to solve a crime.

**3)** *Mystery Festival,* Grades 2-8—This exciting guide features two imaginative and compelling mysteries, one for younger and one for older students.  Students observe the "crime scene" then conduct crime lab tests on the evidence and analyze the results to solve the mystery.

**4)** *Investigating Artifacts: Making Masks, Creating Myths, Exploring Middens,* Kindergarten-6—This guide weaves together three activities related to anthropology and archaeology and to diverse Native American and world cultures.

**National Standards Addressed by this Sequence:**
- Physical Science Standard, K-4: Properties of objects and materials
- Science and Technology Standard, K-4: Abilities of technological design
- History and Nature of Science Standard, K-4: Science as a human endeavor
- Inquiry Standard, K-4: Abilities necessary to do scientific inquiry
- Unifying Concepts and Principles: Evidence, models, and explanation; Constancy, change, and measurement

</td></tr>
</table>

## Grade 5 Physical Science Sequence

Exploring the properties of Materials Through Experimentation and Technology

**1)** ***Bubble-ology,*** Grades 5-9—In this very popular guide, students combine fun and enjoyment with an exploration of important concepts in chemistry and physics through imaginative experiments with soap bubbles.

**2)** ***Bubble Festival,*** Kindergarten-Grade 6—This guide includes 12 classroom table-top activities with set-up instructions. Learning stations encourage independent thinking and cooperative learning—bringing fun and excitement to the classroom, while "bubbling over" with science and math.

**National Standards Addressed by this Sequence:**
- Physical Science Standard, 5-8: Properties and changes of properties in matter
- Science and Technology Standard, 5-8: Abilities of technological design
- Inquiry Standard, 5-8: Abilities necessary to do scientific inquiry
- Unifying Concepts and Principles: Form and function; Constancy, change, and measurement

## Grade 5 Controlled Experiments Sequence

Designing Experiments, from the Simple to the Complex

**1)** ***Hot Water and Warm Homes from Sunlight,*** Grades 4-8—Students build model houses and hot water heaters to discover more about solar power. They conduct experiments to determine the effects of size, color, and number of windows on the amount of heat produced by sunlight.

**2)** ***Paper Towel Testing,*** Grades 5-9—In a series of open-ended experiments, students rank the wet strength and absorbency of four brands of paper towels. Based on their findings and the cost of each brand, they determine which brand is the "best buy."

**3)** ***Vitamin C Testing,*** Grades 4-8—This guide is a stimulating introduction to chemistry and nutrition. The students perform a simple chemical test using a Vitamin C indicator to compare the vitamin C content of different juices and then graph the results.

**National Standards Addressed by this Sequence:**
- History and Nature of Science Standard, 5-8.
- Physical Science Standard, 5-8: Properties and changes of properties in matter
- Earth and Space Science :Standard, 5-8: Transfer of Energy
- Science and Technology Standard, 5-8: Abilities of Technological Design
- Inquiry Standard, 5-8: Abilities necessary to do scientific inquiry
- Unifying Concepts and Principles: evolution and equilibrium; evidence, models, and explanation

<table>
<tr><td valign="top" width="25%">

**Grade 6<br>Earth<br>Science<br>Sequence**

Observing Earth
Materials and
Simulating
Earth Processes

</td><td valign="top">

**1) *Stories in Stone,*** Grades 4-9—Stories in Stone is an earth science unit that deepens understanding of the main processes that lead to the formation of igneous, sedimentary, and metamorphic rocks.

**2) *River Cutters,*** Grades 6-9—Geological time passes very quickly in River Cutters, as students build a model of a river and simulate the creation of a river system in minutes.

**National Standards Addressed by this Sequence:**
- Earth and Space Science Standard, 5-8: Structure of the earth system; Earth's history
- Science and Technology Standard, 5-8: Abilities of technological design
- Science in Personal and Social Perspectives Standard, 5-8: Natural hazards
- Inquiry Standard, 5-8: Abilities necessary to do scientific inquiry
- Unifying Concepts and Principles: Evidence, models, and explanation

</td></tr>
</table>

<table>
<tr><td valign="top" width="25%">

**Grade 6<br>Chemistry<br>Sequence**

Experimenting
with Chemical
Reactions to
Explore Foods,
Everyday
Substances, and the
Environment

</td><td valign="top">

**1) *Chemical Reactions,*** Grades 6-10—An ordinary ziplock bag becomes a safe and spectacular laboratory, as students mix chemicals that bubble, change color, get hot, and produce gas, heat, and an odor.

**3) *Vitamin C Testing,*** Grades 4-8—This guide is a stimulating introduction to chemistry and nutrition. The students perform a simple chemical test using a Vitamin C indicator to compare the vitamin C content of different juices and then graph the results.

**3) *Of Cabbages and Chemistry,*** Grades 4-8—This series of activities offers students a chance to explore acids and bases using the special indicator properties of red cabbage juice.

**4) *Acid Rain,*** Grades 6-10—In this extensive unit that explores many aspects of acid rain and its effects, students gain scientific inquiry skills as they learn about acids and the pH scale, make "fake lakes," determine how the pH of the lakes changes after an acid rainstorm, present a play about the effects of acid rain on aquatic life, and hold a town meeting to discuss solutions to the problem.

**National Standards Addressed by this Sequence:**
- Physical Science Standard, 5-8: Properties and changes of properties in matter
- Earth and Space Science Standard, 5-8: Transfer of Energy
- Inquiry Standard, 5-8: Abilities necessary to do scientific inquiry
- Science in Personal and Social Perspectives Standard 5-8: Populations, Resources, and Environments; Science and Technology in Society
- Unifying Concepts and Principles: Constancy, change and measurement; Evidence, models, and explanation

</td></tr>
</table>

## Grade 6
## Math, Science, and Technology Sequence

Using Geometry and Controlled Experimentation to Investigate Rocket Flight

**1)** *Height-O-Meters,* Grades 6-10—Students are introduced to the principle of triangulation by making simple cardboard devices called "Height-O-Meters." Students measure angles to determine the height of the school flagpole, and compare how high a styrofoam and rubber ball can be thrown. *Height-O-Meters* is a prerequisite for rocketry activities, because it provides students with a practical way to measure the height of a rocket in flight.

**2)** *Experimenting With Model Rockets,* Grades 6-10—Controlled experimentation is introduced in this series of exciting rocketry activities. Students design their own experiments to see how high a rocket will fly by varying the number and placement of fins or the length of the body tube.

**National Standards Addressed by this Sequence:**
* Science and Technology Standard, 5-8: Abilities of technological design; Communicate the process of technological design; Understanding about science and technology
* Inquiry Standard, 5-8: Abilities necessary to do scientific inquiry
* Unifying Concepts and Principles: Evidence, models, and explanation; Constancy, change, and measurement

## Grade 7
## Astronomy and Space Science Sequence

Recording Sky Observations and Constructing Models to Investigate the Solar System

**1)** *Oobleck: What Do Scientists Do?* Grades 4-8—Students investigate and analyze the properties of a strange green substance, Oobleck, said to come from another planet. The class holds a scientific convention to critically discuss scientific findings; and design a spacecraft to land on an ocean of Oobleck.

**2)** *Earth, Moon, and Stars,* Grades 5-8—In this extensive unit, students learn a great deal about the Earth's shape and gravity, moon phases and eclipses, and what the apparent movements of the stars tell us about the earth. Students observe the moon and stars in the sky, create models to explain their observations, and critically discuss their ideas.

**3)** *Moons of Jupiter,* Grades 4-9—In this unit your students become "Galileos" as they re-create, through viewing beautiful slides (provided with the guide) Galileo's historic telescope observations of Jupiter's moons. They also experiment to see how craters are formed, construct a scale model of the Jupiter system, and design a settlement to explore one of Jupiter's moons.

**National Standards Addressed by this Sequence:**
* Earth and Space Science Standard, 5-8: Earth in the Solar System
* Inquiry Standard, 5-8: Abilities necessary to do scientific inquiry
* Science and Technology Standard 5-8: Abilities of technological design; Communicate the process of technological design; Understanding about science and technology
* History and Nature of Science Standard 5-8
* Unifying Concepts and Principles: Constancy, change and measurement; Evidence, models, and explanation; Scale and Structure

<table>
<tr><td valign="top" width="25%">

**Grade 7 Scientific Instruments Sequence**

Experimenting With Light and Color to Discover the Principles that Underlie Scientific Instruments

</td><td valign="top">

**1)** *More Than Magnifiers* Grades 6-10—In this unit, students use the same two lenses in different ways to create optical instruments, and in so doing, find out how lenses are used in magnifiers, simple cameras, telescopes, and slide projectors.

**2)** *Color Analyzers,* Grades 5-9—Students investigate light and color while experimenting with diffraction gratings and color filters. They use color filters to decipher secret messages, then create their own secret messages. By taking part in these activities and drawing conclusions from their experience, students gain further insight into questions such as: Why does an apple look red?

**3)** *Microscopic Explorations,* Grades 4–8—(publication scheduled in 1998). This guide, which explores the many uses of microscopes, features ten learning stations representing a wide spectrum of scientific investigation in many fields of science. Connects well to many other GEMS units, especially the two units listed above.

**National Standards Addressed by this Sequence:**
- Physical Science Standard, 5-8: Transfer of Energy
- Inquiry Standard, 5-8: Abilities necessary to do scientific inquiry
- Science and Technology 5-8: Abilities of technological design
- Unifying Concepts and Principles: Evidence, models, and explanation; Constancy, change, and measurement

</td></tr>
</table>

**Grade 8
Life Science
Sequence**

Making
Systematic
Observations to
Explore Animal
Behavior and
Learning

**1)** *Animals in Action,* Grades 5-9—While watching animals in a large classroom corral, the class adds different stimulus objects to the "corral environment" and observes the animals' responses. Teams of students generate hypotheses, conduct experiments, and hold a scientific convention to discuss their findings. Students learn to distinguish between direct observations and assumptions, as they differential evidence (what they observe) from inferences (conclusions they draw from their observations.)

**2)** *Mapping Animal Movements,* Grades 5-9—Students apply field biology techniques, using a sampling and mapping system, to quantify and compare the movements of hamsters and crickets. Students plan and conduct experiments, graphing changes in movement patterns when food and shelter are added to the environment.

**3)** *Mapping Fish Habitats,* Grades 6-10—Students learn about and apply the field-mapping techniques of aquatic biologists as they chart the movements of fish in a classroom aquarium, and conduct experiments to determine the effects of an environmental change on the home ranges of the fish.

**4)** *Learning About Learning,* Grades 6-8—In this unit, packed with a rich variety of learning modes, students gain insight into the process of their own learning and how learning helps us to survive *and* thrive. Students challenge and coach each other to learn tactile mazes, analyze what helps and hinders learning, and discuss learning differences. They also role-play wolves, simulate actual experiments on how stimulation or deprivation affects rat brain cell growth, and discuss issues of human and animal experimentation.

**National Standards Addressed by this Sequence:**
- Life Science Standard, 5-8: Regulation and Behavior
- Inquiry Standard, 5-8: Abilities necessary to do scientific inquiry
- Unifying Concepts and Principles: Evidence, models, and explanation; Constancy, change, and measurement; Form and function; Evolution

---

**Grade 8
Physical
Science
Sequence**

Investigating the
Interactions of
Matter and
Energy

**1)** *Discovering Density,* Grades 6-10—In a highly involving and colorful hands-on activity, students attempt to layer various liquids in a straw, leading them to explore the concept of density.

**2)** *Convection: A Current Event,* Grades 6-9—Students explore this important physical phenomenon by observing and charting the convection currents in a liquid. Through their own experience, students gain understanding of how patterns they observe in a heated pan represent one of the main ways that heat moves, then go on to apply their understanding to movement of the Earth's crust, the wind and weather, and other natural events.

**National Standards Addressed by this Sequence:**
- Physical Science Standard 5-8: Changes of Properties in Matter; Transfer of Energy
- Earth and Space Science Standard, 5-8: Structure of the earth system
- Inquiry Standard, 5-8: Abilities necessary to do scientific inquiry
- Unifying Concepts and Principles: Constancy, change and measurement; Evidence, models, and explanation

# Chapter 7

# Creating a Local Plan

This handbook has provided an overview of the science education reform movement and shown how the GEMS series in general and GEMS curriculum sequences in particular can be used to fashion science education aligned with the common goals of all the reform documents and in strong support of the National Science Education Standards. The National Standards also strongly emphasize the need for transformation of both science education programs and for the entire science education system. Within this context, the job of planning and implementing a coherent and effective program for your school or district is of central importance. This chapter will provide some suggestions for how you can accomplish the planning task without having to reinvent the wheel. But before we turn to the question of "How?" Let's pause briefly to consider the very important question of "Why?"

## Why Do We Need a Local Plan for Science?

"Why," you may ask, "is it necessary for us to create a plan for our local science program? After all, if state frameworks, *Benchmarks*, and *National Standards* have been written, surely the work must already be done!" This is a good question, and many people deplore what they see as unnecessary redundancy as each of the thousands of school districts in the country "reinvents the wheel" and comes up with its own plan for science. It is very important to recognize that there are valid reasons for local planning, and it is possible to approach the problem in a way that does not require you to do work that has already been done.

Let's start with four compelling reasons for planning your local science program. The first is that one of our great strengths as a nation is that, within broad limits, we have the opportunity and obligation to decide what occurs in our classrooms. Unlike many other countries, in which the curriculum is planned centrally, and all teachers of a given grade level are expected to be on the same page of the same textbook on a given day, we have the opportunity to make decisions about what is best for *our* students. The commitment that comes from the power to participate in those decisions cannot be underestimated. Teachers are much more enthusiastic about a program that they and colleagues whom they respect help to plan.

*...one of our great strengths as a nation is that, within broad limits, we have the opportunity and obligation to decide what occurs in our classrooms,*

A second reason is that a group can change the plan that they created as they see how well it works in practice. There is nothing like experience to tell you if it is really appropriate to teach a particular lesson about light and color at the fifth grade level, or if it is better to postpone it to sixth or seventh grade. Effective, performance-based assessment tasks become enormously useful if the results of those tasks can affect changes in the curriculum plan, so that the science curriculum can evolve over time, and become more educationally effective.

The third reason is concrete and practical.  State and federal guidelines, while increasingly aligned to national standards and benchmarks, tend to be too general to provide sufficient guidance to teachers.  A statement of what a student should know and be able to do by the end of fourth grade does not tell a third grade teacher what unit she should present during the first week of school!  A school-wide or district-wide plan can show how state and federal guidelines can be achieved by taking into account the capabilities and interests of local teachers, available science supplies and equipment, as well as opportunities to visit local museums, planetariums, and other field trip sites.  A local plan will help you to retain and share the best lessons that teachers in your school or district already present, as well as change the aspects of the program that need improvement.  It can help you build on regional experience, adapt to the unique and diverse aspects of your own setting, and be responsive to particular economic, social, and cultural needs and factors.

Last, but far from least, school and district planning makes it possible for teachers to more effectively pool their resources of materials and equipment, to share their expertise, and to collaborate on projects with other teachers.

## Form A Planning Team

The first step is to form a planning team.  In many ways, it is the most critical step of all.  Most school districts create planning teams that consist of a science coordinator,  mentor or resource teachers in science from different grade levels, at least one administrator with responsibility in the area of science and power to affect financial decisions in the district, and several teachers who will be expected to carry out the program in their classrooms.  It is very important that teachers who are selected are not only capable in science, but well-respected by their colleagues.  Community leaders from companies that provide support to the schools and interested parents and scientists who live in the community may be invited as well.

The planning group needs to be large enough to include different points of view, but not too large to obstruct progress.  Sometimes a large group can be broken down into smaller working teams for drafting particular sections of the plan.

Some districts form separate science planning teams for the elementary, middle school, and high school levels.  If more than one team is created, it is important for someone to coordinate the plans of the different teams.  A coordinator might look for opportunities to build in connections and sequences so that students increase their capabilities and understanding of science from year to year by revisiting certain key ideas at higher levels and in different contexts.  A coordinator might also help teams avoid simple repetition of particular lessons at successive grade levels.  Coordination between different grade levels is called *articulation*.  Articulation between the elementary, middle, and high school levels can be one of the most important outcomes of local planning efforts.

Other school districts might form a site planning team at each school.  While that approach has the advantage of building on the strengths and resources at each particular school, the opportunity to join forces with other schools in purchasing science materials and conducting workshops can be lost if there is no coordinator to help the teams share information and look for ways to work together.

## Provide Team Members with a "Tool Kit" for Planning

Once your team is assembled, they will need tools to work with. We urge you to provide each member of your team with a "tool kit" consisting of the following documents.

**All team members should receive:**

- Previous science curriculum guides from your school or district.
- Your state science framework or planning document.
- A copy of this GEMS handbook, *The Architecture of Reform* (1996)
- *Benchmarks for Science Literacy* (1993)
- *National Science Education  Standards* (1996)

Note:  Many of these materials can be purchased from:  National Science Teachers Association, P.O. Box 90477, Washington, DC 20090-0477.  Telephone: (800) 722-NSTA

**Middle and high school planning teams should also receive:**

- *Science for All Americans* (1990)
- *Scope, Sequence and Coordination I. Content Core* (1992)

**When the team is ready to consider assessment, they should also receive:**

- *Insights and Outcomes: Assessments for Great Explorations in Math and Science* (1995)

Arguments are frequently advanced for not providing all of these materials to each planning team member.  It's costly, and the amount of material may seem overwhelming to some.  (Remember the Great Wall of China metaphor.) However, we argue that they should be provided to each team member if at all possible.  First, the cost is small compared with the value of the time that each team member is expected to contribute to the effort.  Second, no one is expected to read these materials at a single setting; they are resources and as such they should be available for consultation when needed.  Third, if we are to avoid reinventing the wheel, everyone needs to be able to take a close look at the wheels that have already been invented!  And finally, as discussed in Chapter 1: The View from the 21st Century, these resources are not simply alternative road maps. Each has a valuable and a complementary role to play in the work that you will do as you evolve a science program for your community.

## Take Stock of Where You Are Now

The more you and your team can learn about the science program currently being presented in your district, and the individuals who are conducting good science programs right now, the more successful your reform efforts will be. Teachers who have worked hard, possibly in the face of adversity or isolation, to create inquiry-based programs need to be acknowledged, and may have ideas and materials to contribute.  These teachers may also play a role in the professional development program, whether or not they are members of the planning team.

In addition to the human resources, take inventory of the material resources in district warehouses and school supply rooms.  In some cases, excellent materials were ordered by people who subsequently left the district, and may still be up-to-date and usable.  Look into films and videos on science, library books and audiotapes.   If possible, make a list of these resources for use by the planning team, and later for use by teachers who want to enrich their science programs.

In addition to the human and material resources within your school district, find out about the local resources for teaching science in your community.  What are good sites for field trips, museums, planetariums, aquariums, botanical gardens, interesting road cuts where students can study geology, or marine environments where students can learn about biodiversity?  Survey parents to see how many might be willing to contribute expertise, time, or materials to assist with improving the science program.  Consider ways to inform and educate all parents about the changes planned in the science program.  Their understanding and support can be a crucial factor in longterm student success.

Before creating a new plan, try to locate any previous plans for science.  Were the plans ever implemented?  Were materials ever ordered?  What parts of the old plans are worth keeping?

Finally, make a list of all major efforts to improve education at your school or district over the past five years, and think of ways that science education reform might build on those efforts even if they seem unrelated at first.  For example, a program to improve reading might be followed by a science initiative that identifies children's literature to accompany each new unit in science.  Or a program to improve mathematics might be followed by new science units which emphasize math skills that have been introduced at the same grade level.  Teachers grow weary of new programs to improve their teaching before they've fully incorporated the last set of reforms; but if they see new initiatives as building on one another, they will be much more willing and able to take on new tasks.


## Plan a Core Science Program, Grade-By-Grade

Following is a set of suggestions for a document that teachers in your school or district are likely to find useful, and how to go about creating it.  They are, of course, only suggestions.  You will need to come up with an overall plan that fits your local needs and requirements.

**Start with a list of strengths and needs in your school or district.**  Start with the information discussed on the previous page—the strengths and gaps in the current program, human and material resources in your community, and the recent history of school reform.  Make sure that whatever is working well now is built in to the new plan that you are about to create.  When you have a good idea of what is available, and what is needed, you're ready to sketch out a science program.  Make sure to build upon flexible ways to weave science across the curriculum, demonstrating its effective use in developing literacy, mathematical reasoning, and language acquisition and development.

**Develop a short statement of principles to guide your work.**  In order to
bring together all of the various elements into a single coherent program, and
make decisions about what science units should be "in" or "out," your team will
need to create a short, useful document that describes the program in broad
strokes.  Do you want to make a  commitment to inquiry-based learning at all age
levels?  Do you want to emphasize cooperative learning?  An integrated
mathematics and science program?  Connections across the curriculum?  Make a
list of the criteria that your team values which reflects both the national reform
documents and the special needs of your community.  You might call it a
"Science Education Philosophy" statement, or a "Statement of Principles" for
designing a new science program.  Be sure to allow some time for input from
people outside the planning group, so that this wider group who have been
consulted will be aware of and tend to support the plan when it is finalized.
Consider their comments carefully and integrate changes as appropriate.
Teachers will be much more willing to implement a plan that they had a hand in
creating than one that is thrust upon them with no opportunity for input.

**Rely on existing statements of content as a point of departure.**  The next
step is to determine what  science content to include at each grade level.  When
you are ready to proceed, it is probably NOT a good idea to start with a blank
page.  Select one of the reform documents to guide the development of your
grade-by-grade curriculum.  A great deal of work went into creating these
summaries of what every student should know and be able to do, and there is no
need to do it all over again.  Alternatively, you might consider Scope and
Sequence charts already designed to accompany inquiry-based programs, such as
the suggested GEMS sequences in Chapter 7 of this handbook.  Unless your local
or state education officer has proclaimed which document to follow, the choice is
up to your group.  Remember that there is a great deal of overlap between the
content portions of these documents.

**Content is more than concepts and theories.**  Suppose, for example, that
you choose to work with the *National Science Education Standards* as your
primary guide.  You will probably spend most of your group discussion time
talking about Chapter 6 Science Content Standards; but do not neglect the other
chapters, concerned with teaching, professional development, assessment,
programs, and the educational system.  And within the Content Chapter, do not
focus all of your effort on the standards related to the life, physical, and Earth and
space sciences.  Budget your time so that you can consider how to meet the other
standards, including unifying concepts and processes, science as inquiry, science
and technology, science in personal and social perspectives, and the history and
nature of science.  It may help to assign one chapter or section to each planning
team member who then becomes the "expert and advocate" with responsibility to
ensure that the section is not neglected.

**Work towards a statement of content to be taught at each grade level.**
You'll notice that none of the reform documents spell out what should be studied
during each grade level.  That is the job of local committees like yours.  It was
not intended for students to achieve all of the standards in one year, but rather
over a span of four or five years.  It is up to you to decide which concepts,
processes, or theories should be revisited two or three times during that period,
and which should be presented only once; which should be presented in first
grade, and which in grade four; and so on.

**Identify core and enrichment units.** A list of concepts and skills is not enough to guide instruction at the local level. Your short document should include a list of science units (including teacher's guides and materials) that already exist in your district, or that can be purchased before the plan is put into place. Putting in names of actual units will clarify how much time each one will take. Stuffing too many units into each grade level will tend to re-create one of the very problems that the reform documents are trying to avoid. Units that are included in the local plan should be considered "core units" if they are to be presented by all teachers to all students at the specified grade level. Time should be allowed for teachers to go further with the core units, or to present additional enrichment units that may be favorites.

**Create a chart showing the core and enrichment units selected for each grade level.** As you develop a plan, try to keep the possibilities fluid for a while. Some people like to create a large wall grid with grade levels forming rows, and general subject areas, like life, physical and earth-space science, forming the columns. A separate list might be posted on the side, showing criteria for all units selected, such as: inquiry-based, real world applications, integrated with mathematics or other subjects, etc. Specific topics can then be written on post-its and moved around on the grid, until your group is pleased with the result. Even if you start with a curriculum plan, such as the GEMS sequence suggested in Chapter 7, it is valuable to create a wall chart showing all of the units to help you consider additions and modifications.

**Incorporate existing elements of the current program.** This is a good time to integrate all of the good things that are already happening in your district, or exciting ideas that teachers have proposed. For example, if the sixth grade teachers already present a wonderful unit on biodiversity using a grassy area and a pond near the school, you will certainly want to include that unit in an overall plan.

**Maintain course and direction.** On the other hand, if most teachers in your school or district only teach science from a textbook, it is probably NOT a good idea to organize the entire plan around the Table of Contents, as it is likely to further reinforce use of the textbook. Instead, focus on finding inquiry-based units that address specific standards, increase student motivation, and provide avenues for independent exploration. Then build the curriculum around the core units.

**Choose science units that meet more than one goal.** In order to keep the number of core units small, each one that is selected will need to achieve several goals. For example, the GEMS unit *Earth, Moon, and Stars* can be used to meet standards with respect to the history and nature of science, as well as astronomy concepts in the Earth and space science standards. The GEMS units can be used to meet standards regarding science as inquiry, in addition to meeting specific content standards in the natural science areas.

**Invite input from all teachers who are expected to implement the program.** When your group has completed a draft, it will be more appropriate, effective, democratic, and diplomatic to use it as a springboard for discussion rather than presenting it as a *fait accompli*. One way to do this is for each member of the planning team to meet with small groups of teachers to discuss the plan and receive feedback. Then, the planning team should meet again and try to incorporate the feedback while still adhering to the basic principles and criteria that had been selected to guide the planning. Even then, you will probably want to consider it a "Working Document" during the first few years of implementation.

## Create a System for Distributing Science Supplies

Inquiry-based science and mathematics programs require materials. Some teachers collect the materials that they need on their own, but it takes time, effort, and a materials budget to do so. One of the great advantages of a district-wide science curriculum plan is to save time and money by coordinating these efforts.

The *GEMS Kit Builder's Handbook* will be of assistance in this regard. The *Handbook* contains a complete illustrated list of all materials that are needed to present each GEMS unit. In addition, many ideas are provided for how to organize science supplies and equipment for any activity-based science program.

If at all possible, obtain at least some of the materials that will be needed before starting a professional development program to support your local plan. If teachers can receive the materials that they will be using during a hands-on workshop, they will be much more likely to use the materials.

A schedule for science should be worked out so that resources can be pooled among groups of teachers at a school or within a district. It also helps to clearly specify in advance how to obtain funds for restocking the kits, who is responsible for replenishing each kit, and how the kits are to be transported from classroom to classroom or school to school. "Surprising" teachers at the first workshop with a list of their responsibilities is probably not a good idea! A more diplomatic approach would be to meet with the teachers to brainstorm all of the tasks that should be done and ask for volunteers. Alternatively, other district personnel may be able to take on some of these jobs, or parent volunteers may be enlisted to help out.

## Establish a Professional Development Program

Given a local curriculum plan, a set of Teacher's Guides, and kits of science materials, some teachers will be able to get started right away. Others may greatly benefit from a professional development program to help get them started. Of course, all teachers need and deserve to frequently take part in such programs to update and deepen their learning and teaching. At their best, such programs should be sustained over a year or more, provide opportunities for in-depth multi-day workshops or summer institutes, and include both presentation of the new units in the classroom and the opportunity to reflect with others on the effectiveness of both the presentation and the unit itself.

The standard model for professional development has been a "one shot" workshop held during a staff development day or even during a faculty meeting. While these can serve an informational and inspirational purpose, for teachers who need assistance, this approach is inadequate. All teachers, especially teachers who are not experienced at teaching inquiry-based science will do best if they can experience a hands-on workshop in *each* unit that they will be expected to teach. To become familiar with a new hands-on science program preferably requires a summer workshop program of at least a week or two in duration. Regardless of time frame, goals, or experience level, the most successful workshops provide direct hands-on experience with the units, rather than quick reviews of several units or short "demonstrations." Teachers learn best by doing and tend to teach the way they are taught. The scope of each workshop should follow the "less is more" principle, so the presentation is immediately useful to participants, not intimidating or overwhelming.

A wide variety of other models and ideas for professional development programs can be found in various sources.  (See, for example, Chapter 4 of the *National Science Education Standards,* 1996, or *Continuing to Learn* by Loucks-Horsley et. al., 1987.)  Creating a plan for professional development should evolve organically once you have a curriculum plan that is supported by the teachers, administrators, and parents in your school or district.

Involving the entire community in the launching and implementation of new curricula can be exciting and effective.  Some schools have used an all-school immersion approach to kick off a new science program. In this approach all teachers at a school attend a one-day workshop on a single core unit, and they are then  expected to present it to their students during a single week.  Units in different classes might be related by topic, such as "Oceans Week," or "Plants and Animals Week," with specially planned events so that students in different classes can show off their work, and the younger students can see what they will have to look forward to in future years.  The GEMS Festival guides, such as *Bubble Festival*, *Build-It Festival*, and *Mystery Festival* been used very effectively in this connection.

Carrying out the plan will undoubtedly require leadership and financial support.  A start-up budget will provide for the initial science supplies and equipment, experts from your own district or from outside who are experienced in using the materials in teacher workshops and with students; and whatever additional funds might be needed to support hands-on workshops (for facilities, teacher stipends or substitutes, lunches, coffee and fruit) so that teachers can focus on the task at hand.

An annual budget will be needed for teachers to meet with each other on a regular basis during the school year to discuss common concerns and share ideas; for master teachers to work with less experienced teachers during summer institutes and during the school day; and for annual assessment, program reviews, and occasional replacement of units with newer more experimental units.

Assistance is also available from sources outside of your school district in most parts of the country.  Some states support Educational Services Districts (ESDs) to provide training, kits, and other support services to schools.  In other states, they're called BOCES or County Offices of Education.

At this writing, more than 25 GEMS Network Sites and Centers exist in various parts of the country to assist schools and districts in their regions that wish to implement an inquiry-based program using GEMS science and mathematics units.  To find out which GEMS Site or Center is nearest you, call GEMS Headquarters at (510) 642-7771.  The GEMS Network and many other national resources and networks have been developed with the same idea in mind—to provide teachers with the consistent and sustained encouragement, regional support, and longterm follow-up that is needed to successfully achieve excellence in science and mathematics education.

## Get Started!

If you wait for full financial support from your School Board, your program will probably never get off the ground!  All you need to get started is a tentative plan, some willing teachers, and enough funds to purchase the first few units.

It is also a good idea to try and involve at least two or three teachers from each school site.  We all need mutual support, and site teams have long been recognized as more effective than isolated teachers, working on their own against the tide.

Similarly, try to enlist the active support of an administrator at the school—the principal or assistant principal in charge of curriculum—and spend enough time with that person so that they have a realistic picture of what to expect.  Ideally, administrators need to participate themselves in activity-based workshops so they can become acquainted with the practice and pedagogy of science education reform.  An administrator who expects quiet students sitting in rows obediently reading their books can discourage teachers from using a hands-on, minds-on approach.

Try to increase the number of teachers who are using the new programs as quickly as possible.  Teachers need to be rewarded—or at least appreciated—for trying something new.  Extra pay for attending Saturday or summertime workshops, or even good snacks during Staff Development Day can go a long way to enlisting the aid of teachers who might have other priorities than science.

## Assess and Revise… Assess and Revise!

Once the new program is underway, it will be important to assess student progress using portfolios, performance tasks, and embedded assessment instruments.  *Insights and Outcomes: Assessments for Great Explorations in Math and Science* will provide you with a lot of ideas for how to do this.  Other excellent programs also have assessment components.

Keep in mind that teachers should be using assessment data in at least two different ways.  One is to provide feedback to their students and their students' parents on how well they are doing in class, and what they might do to improve.  The other is for the teacher to improve the instructional program.

Whenever possible, look at what the students understand about a topic before and  after teaching the unit.  If, in your opinion, there is no significant change in student understanding, the unit has to be changed, or it is probably not worth repeating the next year.

Assessment and revision should be an ongoing process.  A unit that worked well five years ago may not be meeting your students' needs as well as a new unit that just came out.  On the other hand, you can never be sure unless you try it out and collect some data to see how your students' understanding in science changes as a result of the new unit.

## Conclusion

In summary, local planning is a very important task.  It takes time and effort, art and science, perseverance and diplomacy.  Implementing the plan requires even more time and effort.  And finally, science education reform is never finished.  A good program needs to be assessed periodically, and revised whenever it appears that something is not working, or there is a new idea that needs to be tested.

# A Thought in Closing

In this handbook, we have attempted to synthesize a large body of educational thinking in a useful and accessible way. Within the larger framework of educational change, we have sought to demonstrate how the GEMS series can play an important role in achieving "excellence and equity" for all our students. We hope we have helped catalyze your thinking about educational transformation in your own setting.

We are aware that the goals of this handbook are quite ambitious, even audacious. It is not possible in brief summaries to do justice to the leading documents themselves. We urge you to read and ponder these excellent writings for yourself and, in the best inquiry-based tradition, to make up your own mind about their recommendations.

To make matters even more difficult, the worlds of both science and science education are of course in a constant state of change. The rapid expansion of knowledge and technology is certain to produce profound and as yet unanticipated effects on all of our lives as well as in science education. Add to this the lively ferment and controversy that always exists as teachers and educators seek to incorporate new educational and psychological research into their efforts and strive on a daily basis to do more for their students and improve their teaching.

Transformation—so sorely needed in all of education—and most definitely demanded in science, mathematics, and related fields—is complex and multifaceted. As all the reform documents emphasize, the community and society in general must be mobilized for authentic change to occur, for the main elements of the common vision to become reality. National standards and the shared vision for educational excellence are now high on the agenda of educators as well as local, state, and national governments. But such standards and goals—no matter how praiseworthy—can never by themselves transform the daily educational experience of preschoolers, second graders, junior high and high school students.

As Dr. Glenn T. Seaborg said in one of his regular columns in the *GEMS Network News*:

> ...I would not be the experimental scientist I have always been if I did not make several observations about how change takes place. As in the GEMS guide *Chemical Reactions*, change has a way of bubbling up from the bottom. Goals and documents deserve their due, but the achievement of even a portion of the changes they advocate involves far-reaching social transformation...it has to involve people and their daily lives—in the case of our schools and education system, this means parents and teachers, children and older students, administrators, scientists, education professors, community activists and school board members, local and national politicians—all concerned citizens. It involves resources, budgets, and salaries, construction and maintenance, teaching and technology...
>
> Agreed-upon national standards and articulated benchmark-notches on the spreading tree of scientific knowledge may be a necessary step toward the next level of scientific literacy—not, however, if they are viewed as gospel or handed down in a condescending or elitist manner. For at the end of the day, I believe that students and teachers are the real treasures. In some deep way, I believe in each teacher—each one of you—constructing your own standards and benchmarks—as the sum total, always changing, of what your own life and teaching experience have taught you. It is through that experience that you have developed your innate talents—to communicate, to guide by the Socratic question, to coordinate and facilitate, to nudge growing minds onward, to inspire and draw out enthusiasm and curiosity in your students. Through this experience you have gained and given that most precious of human gifts—the ability to teach!

There is no one right way. There are many ways to establish a solid and lasting foundation for your students....to retain their interest and enthusiasm while challenging their minds, setting high expectations, and meeting curriculum goals that align with national standards.

Leading educators nationwide have articulated the goals—the object, however, is to transform these aspirations into reality for all of our students. In the global village of many millions of houses of science, we reap our rewards each time we see a new concept dawning in the eyes of a child.

It is our task to assist the young people of coming generations construct their own houses of science literacy so they are better prepared for the complex problems of the future---and that may be the highest standard to which we can aspire.

## *Students and teachers are the real treasures...*

# References


Barber, Jacqueline, Bergman, Lincoln, and Sneider, Cary, *To Build A House: GEMS and the "Thematic" Approach to Teaching Science*, Berkeley, CA: Lawrence Hall of Science, 1991.

Barber, Jacqueline, Bergman, Lincoln, Goodman, Jan M., Hosoume, Kimi, Lipner, Linda, Sneider, Cary, and Tucker, Laura, *Insights and Outcomes: Assessments for Great Explorations in Math and Science,* Berkeley, CA: Lawrence Hall of Science, 1995.

Brearton, Mary Ann, presentation at the 1996 NSTA Convention in St. Louis, MO, reported that she found 90% agreement between the *Benchmarks for Science Literacy* and the *National Science Education Standards* about what is important for students to know, and what should be excluded from the curriculum, 1996.

Elmore, Richard F., "Getting to Scale with Good Educational Practice," *Harvard Educational Review*, Vol. 66, No. 1, Spring, 1996.

*A Framework for High School Science Education,* Arlington, VA:  National Science Teacher's Association, 1996.

Loucks-Horsley, Susan, Hewson, Peter, Love, Nancy, and Stiles, Katherine (with Hubert Dyasi, Susan Friel, Judith Mumme, Cary Sneider, and Karen Worth)*Designing Professional Develpment for Teachers of Science and Mathematics,* Corwin Press, Thousand Oaks, California, 1997 (in press).

National Commission on Excellence in Education, A Nation At Risk: The Imperative for Educational Reform, Washington, D.C: U.S. Government Printing Office, 1983.

National Research Council, National Committee on Science Education Standards and Assessment, *National Science Education Standards*, Washington, D.C: National Academy Press, 1996.

National Research Council, "Mathematics and Science Education Around the World: What Can We Learn from the Survey of Mathematics and Science Opportunities and the Third International Mathematics and Science Study? Washington, D.C.: National Academy Press, 1996.

*NSTA Awareness Kit for the National Science Education Standards,* Arlington, VA: National Science Teachers' Association, 1996.

Pearsall, M.K. (ed.), Scope, Sequence, and Coordination II. Relevant Research, Washington, D.C:  National Science Teacher's Association, 1992.

Project 2061, American Association for the Advancement of Science, *Benchmarks for Science Literacy*, New York, Oxford: Oxford University Press, 1993.

Project 2061, American Association for the Advancement of Science,"Common Ground: Benchmarks and National Standards," in *2061 Today: Science Literacy for a Changing Future, Newsletter of Project 2061* Volume 5, Number 1, Spring, 1995.

Project 2061, American Association for the Advancement of Science, *Science for All Americans*, New York, Oxford: Oxford University Press, 1990.

*Resources for Science Literacy: Professional Development,* CD-ROM for Windows and Macintosh, Project 2061, American Association for the Advancement of Science, New York, Oxford: Oxford University Press, 1996.

Scope, Sequence, and Coordination of Secondary School Science, *Volume I. The Content Core, A Guide for Curriculum Designers*, Washington, D.C: National Science Teachers Association, 1992.

Sneider, Cary and Ohadi, Mark, "Unraveling Students' Misconceptions About the Earth's Shape and Gravity," submitted to *Science Education,* November, 1995.

California Department of Education, *Science Framework for California Public Schools, Kindergarten Through Grade Twelve*, Sacramento, CA: California State Board of Education, 1990.

# Appendices

The appendices provide greater detail about some of the issues discussed in the text, and should also be helpful in selecting particular GEMS units to bridge gaps in your curriculum plan.

**Appendix A**   compares what each of the three major reform projects have to say about key ideas that are central to the current reform movement.

**Appendix B**   defines ten *themes*, or *unifying concepts and processes* that help students connect different disciplines and topics in science, with examples from the GEMS series. A table "A Guide to Themes in GEMS" displays the major themes found in each GEMS guide.

**Appendix C**   summarizes different aspects of the *nature of science* issues that are featured in GEMS guides. A table displays the categories by GEMS guide.

**Appendix D**   is a chart providing a preliminary look at GEMS guides and National Standards.

**Appendix E**   includes transparency masters that you can use to spark an informative and stimulating discussion at your school or district about science education reform.

# Appendix A

# Comparing the Vision of Three Reform Projects

Chapter 3 identified twelve key ideas that are common to nearly all of the documents that support science education reform.  As a distillation of ideas, it's a valuable list, drawing out the most important principles from a large stack of documents.  However, any distillation of such a rich array of ideas will certainly miss the nuances and shades of meaning and emphases contained in the original sources.

Appendix A is intended to provide the reader with a deeper look into the twelve key ideas from the viewpoint of each of the major reform projects.  Each page is a table which includes a few quotes from the reform documents, supporting the key idea listed at the top.  The quotes do not include all of the information on that topic, but a few sentences that we felt best represented the project's position.

If you find these quotes interesting, by all means refer back to the documents from which they are drawn to see how they fit into the overall context.  Each of the project documents has a unique flow of ideas and emphasis, and is best understood as a whole.

A timeline, showing the duration of each of the projects, and the publication dates of each of the documents is shown on the next page.

# Timeline

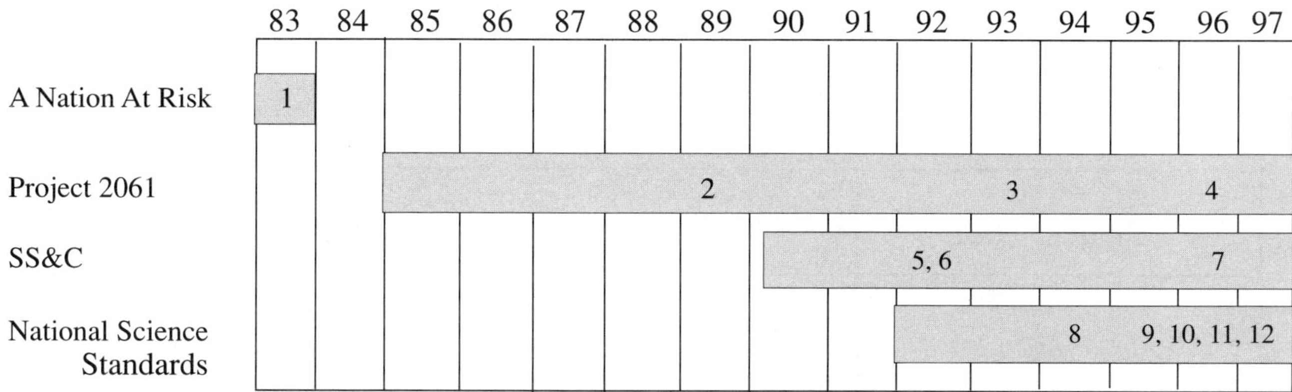



*Shaded bar indicates period of operation of the project.  Numbers refer to publication dates of the following science education reform documents.  These documents are quoted in this Appendix.  Please see the note on page 12 acknowledging the role of the NCTM Standards in this process.*

## National Commission on Excellence in Education

1  National Commission on Excellence in Education, *A Nation At Risk: The Imperative for Educational Reform*, Washington, D.C: U.S. Government Printing Office, 1983.

## Project 2061

2  American Association for the Advancement of Science, Project 2061, *Science for All Americans*, New York, Oxford: Oxford University Press, 1990.

3  American Association for the Advancement of Science, Project 2061, *Benchmarks for Science Literacy*, New York, Oxford: Oxford University Press, 1993.

4 *Resources for Science Literacy: Professional Development*, CD-ROM for Windows and Macintosh, Project 2061, American Association for the Advancement of Science, New York, Oxford: Oxford University Press, 1996.

## Scope, Sequence & Coordination Project

5  Scope, Sequence, and Coordination of Secondary School Science, *Volume I. The Content Core, A Guide for Curriculum Designers*, Washington, D.C: National Science Teacher's Association, 1992.

6  Pearsall, M.K. (ed.), *Scope, Sequence, and Coordination II. Relevant Research*, Washington, D.C:  National Science Teacher's Association, 1992.

7 *A Framework for High School Science Education*, Arlington, VA:  National Science Teacher's Association, 1996.

## National Science Education Standards Project

8  National Committee on Science Education Standards and Assessment, *National Science Education Standards*, *Draft*, Washington, D.C: National Academy Press, 1994.

9  National Committee on Science Education Standards and Assessment, *National Science Education Standards*, Washington, D.C: National Academy Press, 1996.

10 *NSTA Pathways to the Science Standards: High School Edition*, Arlington, VA: National Science Teacher's Association, 1996.

11 *NSTA Awareness Kit for the National Science Education Standards,* Arlington, VA: National Science Teacher's Association, 1996.

12 *NSTA Pathways to the Science Standards: Elementary School Editio*n, Arlington, VA: National Science Teacher's Association, 1997

# 1. Science is for *All* Students

| Project 2061 | SS&C | National Standards |
|---|---|---|
| "*Science for All Americans* consists of a set of recommendations on what understandings and ways of thinking are essential for all citizens in a world shaped by science and technology. (*SFAA*, page v)<br><br>"To reach all students means reforming the education of every strand of the student body—vocational, general, and college preparatory.  For students who expect to go right to work after high school, a narrow focus on trade skills will no longer do; they  need to acquire a strong base of scientific knowledge and of reasoning, communication, and learning skills.  All college-bound students, quite apart from what they believe their majors will eventually turn out to be, need to enter college with an understanding of science, mathematics, and technology that they can build on and that will make it possible for them to elect a technical field.  And undecided students need the knowledge, skills, and attitudes to enable them to move in any direction." (*SFAA,*  page 200) | "...SS&C advocates that instructional strategies be appropriate for heterogenous groups, with no tracking.... Students on all ability levels can exchange ideas and learn from each other... The final goal is ambitious, but not unrealistic: Science learning for all students that is interesting, relevant, challenging, and personally rewarding." (*SS&C Content Core*, pages 15-16) | "The intent of the Standards can be expressed in a single phrase: Science standards for all students.  The phrase embodies both excellence and equity.  The Standards apply to all students, regardless of age, gender, cultural or ethnic background, disabilities, aspirations, or interest and motivation in science.  Different students will achieve understanding in different ways, and different students will achieve different degrees of depth and breadth of understanding depending on interest, ability, and context.  But all students can develop the knowledge and skills described in the Standards, even as some students go well beyond these levels."   (*NSES*, page 2) |

# 2. Commitment to Diversity and Equity

| Project 2061 | SS&C | National Standards |
| --- | --- | --- |
| "The set of recommendations constitutes a common core of learning in science, mathematics, and technology for all young people, regardless of their social circumstances and career aspirations.  In particular, the recommendations pertain to those who in the past have largely been bypassed in science and mathematics education: ethnic and language minorities and girls." (*SFAA*, page x) | "SS&C believes that a diverse classroom improves learning." (*SS&C Content Core*, page 15) | "All students, regardless of sex, cultural or ethnic background, physical or learning disabilities, future aspirations, or interest in science, should have the opportunity to attain high levels of scientific literacy.  By adopting this principle of equity and excellence, the Standards prescribe the inclusion of all students in challenging science learning opportunities and define a high level of understanding that all students should achieve.  In particular, the commitment to science for all implies inclusion of those who traditionally have not received encouragement and opportunity to pursue science—women and girls, students of color, students with limited English proficiency.  It implies attention to various styles of learning, adaptations to meet the needs of special students and different sources of motivation.  And it also implies providing opportunities for those students interested in and capable of moving beyond the basic program." (*NSES*, page 221) |

# 3. Start with a Foundation of Inquiry Skills

| Project 2061 | SS&C | National Standards |
| --- | --- | --- |

**Project 2061**

"Students need to have many and varied opportunities for collecting, sorting and cataloging; observing, note taking and sketching; interviewing, polling, and surveying; and using hand lenses, microscopes, thermometers, cameras, and other common instruments. They should dissect; measure, count, graph, and compute; explore the chemical properties of common substances; plant and cultivate; and systematically observe the social behavior of humans and other animals... Students should be given problems—at levels appropriate to their maturity— that require them to decide what evidence is relevant and to offer their own interpretations of what the evidence means..." (*SFAA,* page 188)

**SS&C**

"Student-centered lessons, with emphasis on hands-on activities, are integral to the SS&C approach. Science programs should help students to answer the questions of science, not by presenting assertions or authority-determined answers, but by allowing them to propose and pursue the ideas, concepts, and information. Teachers should be encouraged to ask: How do we know? Why do we believe? What does it mean?" (*SS&C Content Core,* pages 15-16)

"Science programs should help students to answer the questions of science, not by presenting assertions or authority-determined answers, but by allowing them to propose and pursue ideas, concepts, and information. Teachers should encourage students to ask:

• How do we know?

• Why do we believe?

• What does it mean?"

(*SS&C Content Core,* pages 15-16)

**National Standards**

"Learning science is something students do, not something that is done to them. In learning science, students describe objects and events, ask questions, acquire knowledge, construct explanations of natural phenomena, test those explanations in many different ways, and communicate their ideas to others.

"In the *National Science Education Standards,* the term 'active process' implies physical and mental activity. Hands-on activities are not enough—students also must have 'minds-on' experiences. Science teaching must involve students in inquiry-oriented investigations in which they interact with their teachers and peers. Students establish connections between their current knowledge of science and the scientific knowledge found in many sources; they apply science content to new questions; they engage in problem solving, planning, decision-making, and group discussions; and they experience assessments that are consistent with an active approach to learning." (*NSES,* page 20)

# 4. Construct a Framework of Unifying Concepts and Processes

| Project 2061 | SS&C | National Standards |
| --- | --- | --- |
| "Some important themes pervade science, mathematics, and technology and appear over and over again, whether we are looking at an ancient civilization, the human body, or a comet. They are ideas that transcend disciplinary boundaries and prove fruitful in explanation, in theory, in observation, and in design." (*SFAA*, page 155)<br><br>*SFAA* (pages 155-169) identifies the following themes:<br><br>• Systems<br><br>• Models<br><br>• Constancy<br><br>• Patterns of Change<br><br>• Scale | "A course organized around a Great Idea (or ideas) of Science necessarily integrates those disciplines that operate under its laws or principles.  The notion of 'Great Ideas' suggests that scientists arrived at a single, encompassing idea after making extensive observations and working through  complex chains of thought.  Therefore, course designers must construct Great Idea courses carefully to connect the experiences of children with the larger concept." (*SS&C Content Core*, page 19)<br><br>Examples of courses organized around Great Ideas include: Energy and Evolution. (*SS& C Content Core*, pages 20-21) | "Conceptual and procedural schemes unify science disciplines and provide students with powerful ideas to help them understand the natural world." (*NSES*, page 104)<br><br>"The conceptual and procedural schemes in this standard provide students with productive and insightful ways of thinking about and integrating a range of basic ideas that explain the natural and designed world.<br><br>"As a result of activities in grades K-12, all students should develop understanding and abilities aligned with the following concepts and processes:<br><br>• Systems, order, and organization<br><br>• Evidence, models, and  explanation<br><br>• Constancy, change, and  measurement<br><br>• Evolution and equilibrium<br><br>• Form and function"<br><br>(*NSES*, page 115) |

# 5. Identify the Most Important Facts. Concepts, and Theories

| Project 2061 | SS&C | National Standards |
| --- | --- | --- |

**Project 2061**

"The present curricula in science and mathematics are overstuffed and under-nourished. Over the decades, they have grown with little restraint, thereby overwhelming teachers and students and making it difficult for them to keep track of what science, mathematics, and technology is truly essential... Our fundamental premise is that schools do not need to be asked to teach more and more content, but rather to focus on what is essential to scientific literacy and to teach it more effectively."

(*SFAA*, pages viii, ix)

**SS&C**

"...a consensus is developing among science educators that 'less is more,' i.e., that by covering fewer topics, students can develop a deeper understanding of science. If students learn a few concepts in depth, they can apply them to new situations or problems." (*SS&C Content Core*, pages 14-15)

**National Standards**

"Emphasizing active science learning means shifting emphasis away from teachers presenting information and covering science topics. The perceived need to include all the topics, vocabulary, and information in textbooks is in direct conflict with the central goal of having students learn scientific knowledge with understanding." (*NSES*, page 20-21)

"If teachers are to teach for understanding as described in the content standards, then coverage of great amounts of trivial, unconnected information must be eliminated from the curriculum." (*NSES*, page 213)

# 6. Integrate Ideas from Different Fields Whenever Possible

| Project 2061 | SS&C | National Standards |
| --- | --- | --- |
| "Piecemeal reform measures beget piecemeal effects, if any. At the school district level, reform efforts should be inclusive: all grades, all subject domains, all streams. It is less demanding to concentrate on, say, improving third-grade reading, junior high school social studies, and biology for vocational students. But such unrelated changes are not likely to add up to curricula that are any more integrated, coherent, and effective than the fragmented, overburdened ones that now exist. Without a more sweeping approach, change will be constrained by having to fit within the boundaries of class periods, school subjects, sequences, and tracks that themselves may be a large part of the problem." (*SFAA*, page 199) | "Biology, chemistry, Earth/space science, and physics share topics and processes. Coordination among these four disciplines leads students to an awareness of the interdependence of the sciences and their place in the larger body of human knowledge." (*SS&C Content Core*, page 15)<br><br>"Coordination can be achieved by designing and teaching a single, integrated course in science." (*SS&C Content Core*, page 19) | "Multidisciplinary perspectives also increase from the subject-matter standards to the standard on the history and nature of science, providing many opportunities for integrated approaches to science teaching." (*NSES*, page 104)<br><br>"Student achievement in science and in other school subjects such as social studies, language arts, and technology is enhanced by coordination between and among the science program and other programs. Furthermore, such coordination can make maximal use of time in a crowded school schedule.... Coordination of science and mathematics programs provides an opportunity to advance instruction in science beyond the purely descriptive. Students gathering data in a science investigation should use tools of data analysis to organize these data and to formulate hypotheses for further testing.... At the school level, teachers of mathematics and science must develop and implement a coordinated program." (*NSES*, pages 214-218) |

# 7. Include Examples Illustrating the History and Nature of Science

| Project 2061 | SS&C | National Standards |
|---|---|---|
| "When people know how scientists go about their work and reach scientific conclusions, and what the limitations of such conclusions are, they are more likely to react thoughtfully to scientific claims and less likely to reject them out of hand or accept them uncritically.<br><br>"Once people gain a good sense of how science operates—along with a basic inventory of key science concepts as a basis for learning more later—they can follow the science adventure story as it plays out during their lifetimes.<br><br>"The images that many people have of science and how it works are often distorted. The myths and stereotypes that young people have about science are not dispelled when science teaching focuses narrowly on the laws, concepts, and theories of science. Hence, the study of science as a way of knowing needs to be made explicit in the curriculum." (*Benchmarks*, page 3) | "Biology, chemistry, Earth/space science, and physics share topics and processes. Coordination among these four disciplines leads students to an awareness of the interdependence of the sciences and their place in the larger body of human knowledge. (*SS&C Content Core*, page 15)<br><br>"...curricula should present science as first and foremost a way of (1) learning about the behavior of the universe and the matter and energy it contains; (2) organizing this knowledge so that it is comprehensible and useful to humans; and (3) developing models and theories about this behavior that not only correlates with past observations but also helps predict future events." (*SS&C Content Core*, page 19) | "In learning science, students need to understand that science reflects its history and is an ongoing, changing enterprise. The standards for the history and nature of science recommend the use of history in school science programs to clarify different aspects of scientific inquiry, the human aspects of science, and the role that science has played in the development of various cultures." (*NSES*, page 107)<br><br>"Experiences in which students actually engage in scientific investigations provide the background for developing an understanding of the nature of scientific inquiry, and will also provide a foundation for appreciating the history of science described in this standard... The introduction of historical examples will help students see the scientific enterprise as more philosophical, social, and human." (*NSES*, page 170) |

# 8. Show How Science and Technology Inform Societal Issues

| Project 2061 | SS&C | National Standards |
|---|---|---|
| "More and more, citizens are called on to decide which technologies to develop, which to use, and how to use them.  Part of being prepared for that responsibility is knowing about how technology works, including its alternatives, benefits, risks, and limitations.  The long-term interests of society are best served when key issues concerning proposals to introduce or curtail technology are addressed before final decisions are made.  Students should learn how to ask important questions about the immediate and long-range impacts that technological innovations and the elimination of existing technologies are likely to have.  But intelligent adults disagree about wise use of technology.  Schooling should help students learn how to think critically about technology issues, not what to think about them.  Teachers can help students acquire informed attitudes on the various technologies and their social, cultural, economic, and ecological consequences." (*Benchmarks*, page 53) | "An SS&C curriculum... incorporates practical applications of science—engaging students in the early grades with science problems and issues of personal concern, and in the later grades with more global science considerations.  Gradually putting science in a larger context helps students relate science to themselves and their lives. (*SS&C Content Core*, page 15)) <br><br> "A Science, Technology, and Society course integrates disciplines under the rubric of science in its modern context.  The challenge for those designing STS courses is not finding appropriate issues, but rather selecting a set of issues that encompasses all of the science concepts included in The Content Core.... Certain STS courses would require students to handle more complex science knowledge and experiences and to sort out the competing claims of science and society.  For instance, school districts that lie east of heavily industrialized regions might design a single, integrated course around acid rain... Similar to acid rain, global warming could act as a course focus, but its conflicting data would need to be as carefully presented as its scientific premises." (*SS&C Content Core*, pages 22-23) | "An important purpose of science education is to give students a means to understand and act on personal and social issues.  The science in personal and social perspectives standards... give students a foundation on which to base decisions they will face as citizens." (*NSES*, page 107) <br><br> "[The science in personal and social perspectives standard] includes issues such as personal health, human population growth, natural resources, environmental change, natural and human-induced hazards, and science and technology in local, national, and global challenges." (*NSES*, page 108) |

# 9. Each Student Must Construct His or Her Own House of Science

| Project 2061 | SS&C | National Standards |
| --- | --- | --- |
| "But effective learning often requires more than just making multiple connections of new ideas to old ones; it sometimes requires that people restructure their thinking radically.... Students come to school with their own ideas, some correct and some not, about almost every topic they are likely to encounter.  If their intuition and misconceptions are ignored or dismissed out of hand, their original beliefs are likely to win out in the long run, even though they may give the test answers their teachers want... students must be encouraged to develop new views by seeing how such views help them make better sense of the world." (*SFAA*, page 186) | "Since experiences precede the mastery of terminology, a constructivist approach to learning that responds to student preconceptions is most appropriate." (*SS&C Content Core*, page 15) | "Analysis of student assessment data provides teachers with knowledge to meet the needs of each student.  It gives them indicators of each student's current understanding, the nature of each student's thinking, and the origin of what each knows.  This knowledge leads to decisions about individual teacher-student interactions, to modifications of learning activities to meet diverse student needs and learning approaches, and to the design of learning activities that build from student experience, culture, and prior understanding." (*NSES*, page 42) |

# 10. Students Collaborate in Teams Much of the Time

| Project 2061 | SS&C | National Standards |
|---|---|---|
| "The collaborative nature of scientific and technological work should be strongly reinforced by frequent group activity in the classroom. Scientists and engineers work mostly in groups and less often as isolated investigators. Similarly, students should gain experience sharing responsibility for learning with each other. In the process of coming to common understandings, students in a group must frequently inform each other about procedures and meanings, argue over findings, and assess how the task is progressing. In the context of team responsibility, feedback and communication become more realistic and of a character very different from the usual individualistic textbook-homework-recitation approach." (*SFAA*, page 189) | "Students on all ability levels can exchange ideas and learn from each other." (SS&C Content Core, page 15) | "Science is often a collaborative endeavor, and all science depends on the ultimate sharing and debating of ideas. When carefully guided by teachers to ensure full participation by all, interactions among individuals and groups in the classroom can be vital in deepening the understanding of scientific concepts and the nature of scientific endeavors... Teachers of science must decide when and for what purpose to use whole-class instruction, small-group collaboration, and individual work." (NSES, pages 31-32)<br><br>"Working collaboratively with others not only enhances the understanding of science, it also fosters the practice of many of the skills, attitudes, and values that characterize science. Effective teachers design many of the activities for learning science to require group work, not simply as an exercise, but as essential to the inquiry. The teacher's role is to structure the groups and to teach students the skills that are needed to work together." (*NSES*, page 50) |

# 11. Align Assessment with Curriculum and Teaching

| Project 2061 | SS&C | National Standards |
|---|---|---|
| "Project 2061 has convened expert groups to prepare, in collaboration with the six Project 2061 teams, a dozen concept papers on aspects of the system that must change to accommodate the curriculum reforms being proposed by the project. The findings of these papers... will be presented in *Blueprints for Reform.* The Blueprint topics... [include] Assessment [which] will specify what immediate and future assessment needs are demanded by Project 2061 curriculum design principles—from in-class assessment during instruction, to program evaluation by schools, to monitoring education progress at state and national levels." (*Benchmarks*, page 381-382) | "...students' progress cannot be assessed unless assessment instruments are consistent with what students are expected to learn." (*SS&C Content Core*, page 29) | "Assessment practices and policies provide operational definitions of what is important. For example, the use of an extended inquiry for an assessment task signals what students are to learn, how teachers are to teach, and where resources are to be allocated." (*NSES*, page 76)<br><br>"Another important shift is toward 'authentic assessment.' This movement calls for exercises that closely approximate the intended outcomes of science education. Authentic assessment exercises require students to apply scientific knowledge and reasoning to situations similar to those they will encounter in the world outside the classroom, as well as to situations that approximate how scientists do their work." (*NSES, page* 78)<br><br>"...the alignment of assessment with curriculum and teaching is one of the most critical pieces of science education reform. If the assessment system at the school and district levels does not reflect the Standards and measure what is valued, the likelihood of reform is greatly diminished." (*NSES*, page 211) |

# 12. Mobilize the Whole System to Support the New Reforms

| Project 2061 | SS&C | National Standards |
|---|---|---|
| "Nationwide, reform needs to be comprehensive in the sense of addressing all aspects of the system. Reform in science education depends on changing existing curricula from kindergarten through high school.  But to make the new curricula work, changes must also occur in the preparation of teachers, the content of textbooks and other learning materials, the use of technologies, the nature of testing, and the organization of schools. Furthermore, the changes need to be compatible, lest they cancel each other out." (*SFAA,* page 199) | "...designing and implementing an integrated and coordinated science program is a complex task.  Advocates for restructuring must talk with state legislatures and educators.  As publishers reassess the needs of the market, curriculum designers and teachers may need to take the initiative to compile and/or develop appropriate SS&C materials.... Scientists, educators, teachers, parents, and community members must join together to support science education restructuring efforts." (*SS&C Content Core*, page 16) | "The educational system must act to sustain effective teaching.  The routines, rewards, structures, and expectations of the system must endorse the vision of science teaching portrayed by the *Standards*.  Teachers must be provided with resources, time, and opportunities to make change as described in the program and system standards.  They must work within a framework that encourages their efforts." (*NSES*, page 28)<br><br>"The *Standards* took an insightful and innovative step by suggesting that the responsibility for improving scientific literacy extends beyond those in classrooms and schools to the entire educational system..... All of the science education community—curriculum developers, superintendents, supervisors, policy makers, assessment specialists, scientists, teacher educators—must act to make the vision of these standards a reality."    (*NSES*, page 244) |

# Appendix B
# Major Themes Defined

Here are definitions of each major theme that we have selected to provide a framework for the GEMS program. Our list of themes is consistent with similar lists in *Science for All Americans* and the *National Science Education Standards*. However, just as there is no single scientific method, *no* list of themes is intended to be all-inclusive or authoritative. We provide a few examples from the GEMS series to illustrate the meanings of these big ideas; but the examples are not meant to be exhaustive. Many other GEMS units can also be related to a particular theme or themes.

## Systems and Interactions

A **system** is any collection of things that have some influence on one another and appear to constitute a unified whole. In defining a system, enough parts must be included so the relationship of the parts to each other makes sense. Where we draw the boundaries of any system makes a difference in terms of understanding what is going on. Any part of a system also can be considered as a system or subsystem, with its own internal parts and interactions. Systems often overlap and are not mutually exclusive. Change in the operation or structure of any part of a system has an impact on the rest. Analyzing the input and output of a system and investigating instances of feedback are also important to understanding systems.

Students investigate the circulatory system in *Earthworms*. They observe the vessels and hearts of an earthworm and count the pulses of blood that course through the worm. They consider both the circulatory system of the earthworm and how it interacts with changes in the environment.

*Buzzing a Hive* involves students in investigating the system of a honeybee hive. Students discover the parts of a hive and how they interact. They also explore the interactions of bees and flowers and how this results in both pollination of flowers and collection of an important food stuff for the honeybee.

## Models and Simulations

A **model** is defined as a simplified imitation that we use to better understand a phenomenon, system, or process. The model may be a device, a plan, a drawing, an equation, a computer program, or a mental image. The value of models lies in suggesting how things work or might work. Models, be they physical, conceptual, mathematical, or computer-based, have served invaluable functions in enhancing human scientific understanding. However, models can also be misleading, suggesting characteristics, mechanisms, or analogies that do not exist or are inaccurate, so the limitations of models also need to be understood. A dynamic model is usually referred to as a **simulation**. Simulations represent natural processes in the form of, for example, a game, computer program, or dramatization.

The rivers and geological features that students create in *River Cutters* are examples of model rivers. This particular model river system enables students to simulate the passage of time as they create a speeded-up model of how our planet is carved by water. Students introduce dams and toxic waste dumps in their model river systems, thereby simulating the impact of each of these human interventions on a river and its surrounding area.

In *Hide a Butterfly*, students make and use paper models of flowers, butterflies, and birds as they enact the drama of a butterfly escaping the notice of a paper bag bird by blending into the colors of the flowers. This playful simulation of protective coloration is one of many activities in GEMS in which young children make models and simulate natural relationships.

## Stability

**Stability** (or constancy) represents the ways in which systems do **not** change. Most physical systems settle into a state of **equilibrium** at a particular point, in which the forces within it are balanced until something new is done to the system. Many systems include feedback subsystems that serve to keep some aspect of the system constant. Whatever changes may happen inside a system, there are some total quantities that remain the same, or constant.

Stability is related to the repeatability of results in science and the ability to derive "laws" of physical behavior. Given exactly the same experimental conditions, with no uncontrolled variables, results are expected to be repeated. Similarly, the same law of gravity applies not only to bringing a pop fly back to the ball field, but also to keeping the moon in its orbit, and the mutual attraction between galaxies billions of light years away. The law of gravity is only one of many such principles, in all fields of science, that seem to be constant and unchanging. Without stability of this sort, causality and predictability would have no meaning and science as we know it would be impossible.

In *Mapping Fish Habitats*, students may discover that certain fish always remain at the bottom of the tank, while others spend most of their time in the middle, or at the top. This behavior is predictable and doesn't change even when different factors are introduced to the tank.

Another kind of stability is **conservation**. While many changes occur in the ziplock bag experiment of *Chemical Reactions*, students observe that the bag of reactants has the same mass before and after the chemicals react. This demonstration of conservation of mass, or conservation of quantities, provides a good glimpse of stability in physical systems.

**Static equilibrium** is represented by a rock at the bottom of a hill. Its position remains unchanged because the force of gravity is balanced by the force of the ground underneath the rock. **Dynamic equilibrium** is represented by a person walking down an escalator at the same rate the automatic stairway is moving up. That is, the person's position remains unchanged as long as the downward velocity is equal and opposite to the upward velocity. An example of dynamic equilibrium can be seen in the activities of *Hot Water and Warm Homes from Sunlight*. Students discover that when they place their model houses in sunlight the temperature rises rapidly until the heat gain due to the sunlight balances the heat loss to the surroundings. This is called the **equilibrium temperature.**

## Patterns of Change

Analyzing **patterns of change** is of great importance in the sciences, and central to mathematics as well.  Patterns of change tell us about the relationship between variables and the properties of equations.

Change can be classified into three general patterns: (1) changes that are steady, predictable trends, such as the increasing speed of a falling rock or the rate of radioactive decay; (2) changes that are cyclical, occurring in a sequence that happens over and over, such as the seasonal cycles of weather or the vibration of a guitar string; and (3) changes that appear to be random and unpredictable, or so complicated as to have apparently chaotic results.

For example, in the GEMS guide *Global Warming and the Greenhouse Effect,* students study data on the concentration of carbon dioxide in the Earth's atmosphere.  They can see the annual increase and decrease indicating that huge amounts of carbon dioxide are absorbed by plants during the growing season, and released when plants die and decay in the fall and winter.  But another aspect of the pattern is the gradual increase in the level of carbon dioxide from one year to the next, due to the burning of fossil fuels by people around the world.  Thus, the concentration of carbon dioxide displays both cyclic change and a long term trend.

## Evolution

The main idea of **evolution** (used here in its most general sense, as well as its usual biological reference) is that the present arises, or evolves, from the past.

Our hands, the fins of whales, the wings of bats, and the paws of cats all appear to have evolved from a set of bones in an ancient reptile, through an evolutionary process involving natural selection and numerous other factors.  While many people think of evolution as gradual, the process can include sudden leaps, and even explosions.

Even when a theme, such as evolution, may be too sophisticated for a certain group of students, exposing students to ideas through concrete activities helps lay the groundwork for fuller understanding in subsequent years.  For example, in *Animal Defenses* young students dramatize and discuss the structural and behavioral defenses that "defenseless animals" have evolved to protect themselves from a Tyrannosaurus rex. In *Hide a Butterfly* students explore protective coloration, part of the natural selection process.

In *Involving Dissolving* students observe the changes that occur over time as an egg shell sits in vinegar. While many instances of dissolving are rapid, there are cases, such as this one, in which dissolving occurs at a slow rate.

## Scale

**Scale** concerns the range, comparison, and measurement of various magnitudes, such as sizes, speeds, distances, or lengths of time. While our brains cannot fully grasp the sheer magnitude of the speed of light, we can represent it symbolically, in mathematical terms, and thus further explore the phenomena and its relation to other parts of the physical universe. In some thematic listings, scale is linked with structure. Scale is relative: an ant finds the distance around a tree to be much greater than a squirrel does; a mound of earth to us may be a small mountain to the squirrel.

Students examine small convection currents that occur in a pan of hot water at their desks in *Convection: A Current Event.* They go on to speculate on the effect of enormous convection currents on  global weather patterns, on the movement of entire continents, and on huge masses of material within the sun and other stars.

In *Height-O-Meters* students measure the angular height of a flagpole, learning the same triangulation process (based on the similarity of triangles no matter how large or small they may be) that astronomers use to measure the distances to stars.

In *Moons of Jupiter* students create a model of the Jupiter system on the playground.  In contrast to the tiny scale of the Earth and its moon, it is apparent that the Jupiter system is vastly larger.

## Structure

**Structure** refers to the way in which things are put together, whether it be the hexagonal shapes used in a beehive or an airplane's wing, the internal configuration of an atom, or the geologic structure of the Earth. There is an immense diversity of structure in the natural world. Structure is directly related to both function and to scale. A 50-story skyscraper cannot be constructed in exactly the same way as a two-story house.  A beam that is twice as big is four times as strong, and weights eight times as much. Structure, like scale, can have a central mathematical component.

In *Build It! Festival* student teams build towers out of newspaper dowels, using their knowledge of shapes to make it stable.

In *Buzzing a Hive*  young students investigate the structures of a honeybee, a flower, and a honeybee hive, as they make paper models of each.

In *Fingerprinting* students examine the structure of their fingerprints. They go on to develop their own classification systems to sort the various fingerprint patterns they observed.

In *Animal Defenses* young students create defensive structures to protect their "defenseless" animals. They later discuss the difference between defensive "structures" and "behaviors."

In *Stories in Stone,* students observe the beautiful patterns formed by minerals as they crystallize.  They then apply what they learned as they use magnifiers to examine the structure of a rock for clues to how it was formed.

In *Learning About Learning,* students find out how the structure of neurons enable their brains to make connections that we call "learning."

## Energy

All physical phenomena and interactions involve **energy**. In physical terms, energy is the capacity to do work or the ability to make things move; in chemical terms, it is the basis for reactions between compounds; and in biological terms energy gives living systems the ability to live, grow, and reproduce. One of the most important laws in the physical sciences is that energy can never be created nor destroyed, but only transformed from one form to another. Heat energy can be transferred from one place to another by conduction, convection, or radiation. Energy is very important in Earth science, and the uses and availability of natural resources for the production of energy for human consumption has assumed an increasingly significant role in social issues and world events.

*Convection: A Current Event* enables students to observe convection currents firsthand and to extend the concept to better understand ocean currents, winds, the movement of tectonic plates, and heat transfer within the sun.

Students build model houses and hot water heaters to discover more about solar energy in *Hot Water and Warm Homes from Sunlight*. They conduct experiments to determine the effects of size, color, and number of windows on the amount of heat produced from sunlight.

In *Color Analyzers*, students investigate light by experimenting with filters and discussing how our eyes perceive color.

The way that solar energy is selectively absorbed by carbon dioxide and other greenhouse gases in our atmosphere is a major topic in *Global Warming and the Greenhouse Effect*.

The Assembly Presenter's Guide, *The "Magic" of Electricity* describes a series of impressive demonstrations that provide the audience with the knowledge that electricity can be generated in a variety of ways, through the transformation of other kinds of energy: burning of fossil fuels, nuclear energy, wind, and solar radiation. A model of electricity is presented as moving electrons.

The collection of GEMS Exhibit Guides in *The Wizard's Lab*, provide a variety of ideas for table top exhibits that present students with opportunities to learn about energy by experimenting with pendula, solar cells, light polarizers, magnets, a spinning platform, and electricity makers.

## Matter

**Matter** is the "stuff" that makes up all of the objects that we can observe in the universe, including ourselves. As with energy, the notion of matter is important in all the sciences. Modern physicists and chemists conduct experiments and develop mathematical models to explain the behavior of the smallest known particles of matter at very high energies. Materials engineers study the interactions of various elements and compounds under a wide range of conditions, and try to develop materials with special properties such as superconductivity or high strength. Biochemists try to unravel the material basis for life processes, and geneticists have begun to manipulate the genetic code so that they can someday cure diseases caused by genetic damage.

*Liquid Explorations* lets young students investigate the properties of liquids through a series of engaging activities in which they observe and compare liquids to see the ways in which liquids are similar and the ways in which they are different.

Students investigate the characteristics of different paper towels in *Paper Towel Testing*. By designing experiments to determine the wet strength and absorbency of paper towels, students grapple with the problem of how consumer scientists identify the most important qualities in a product, how to treat the various brands of products fairly, and how to describe the properties of the materials so it makes sense to the potential user.

*In Discovering Density*: students experiment with liquids of various densities, thus enabling them to learn about an important property of matter.

In the Assembly Presenter's Guide *Solids, Liquids, and Gases* the presenter helps the audience hone their definitions of the phases of matter through a number of exciting phase change demonstrations.

*Of Cabbages and Chemistry* enables students to discover another classification system for matter—those substances which are acids, those which are bases, and those which are neither acid nor base, called neutral. Students experiment with different liquids, using the natural acid/base indicator properties of red cabbage juice.

## Diversity and Unity

The theme of **"Diversity and Unity"** represents a way to comprehend and appreciate both the differences and similarities that are interwoven in the sciences and in our understanding of nature and the physical world. From William Blake's "universe in a grain of sand," to the search for a theory that reflects the unity between molecular structure and the dynamics of the universe itself, this idea runs through many fields of human endeavor and thought. The variety of organisms is immense, yet all share many basic processes involved in being alive, and many of these diverse species may be linked and interdependent, and in that way unified, within a larger ecosystem. The immense diversity of living forms, of geological structures, of materials and processes, can be overwhelming. In the early stages of any new science, the first order of business is to begin to classify and categorize. As our practical understanding deepens, unifying ideas develop that may eventually lead to experiments, hypotheses, and theories.

In *Fingerprinting*, students invent systems for classifying fingerprints, and to discover how all humans have similar patterns on their fingertips, but that no one person's fingerprints are exactly the same as another's.

*Animals in Action* : students investigate the similarities and differences between animal behaviors as they conduct stimulus/response experiments with a variety of common animals. All of the other GEMS activities that relate to animal life reflect the theme of "Diversity and Unity" in various ways.

In *Investigating Artifacts* students explore the diversity of cultures and the unity of humankind through the study of mask and myths.

In *Learning About Learning*, students learn about the multiple intelligences that come together in a single human mind.

# A GUIDE TO THEMES

| | Systems and Interactions | Stability | Diversity and Unity | Models and Simulations | Matter | Patterns of Change | Energy | Evolution | Structure | Scale |
|---|---|---|---|---|---|---|---|---|---|---|
| Acid Rain | ✓ | ✓ | ✓ | ✓ | ✓ | ✓ | ✓ | ✓ | | ✓ |
| Animal Defenses | ✓ | | ✓ | ✓ | | | | ✓ | ✓ | ✓ |
| Animals in Action | ✓ | ✓ | ✓ | | | ✓ | | | ✓ | ✓ |
| Ant Homes Under the Ground | ✓ | | ✓ | ✓ | | ✓ | | | ✓ | ✓ |
| Bubble Festival | ✓ | ✓ | ✓ | ✓ | ✓ | ✓ | ✓ | | ✓ | ✓ |
| Bubble-ology | ✓ | ✓ | ✓ | | ✓ | ✓ | | | ✓ | ✓ |
| Build It! Festival | | | ✓ | ✓ | | | | | ✓ | ✓ |
| Buzzing A Hive | ✓ | ✓ | ✓ | ✓ | ✓ | ✓ | ✓ | ✓ | ✓ | ✓ |
| Chemical Reactions | ✓ | ✓ | | ✓ | ✓ | ✓ | | | | ✓ |
| Color Analyzers | ✓ | ✓ | | | ✓ | | ✓ | | | ✓ |
| Convection:  A Current Event | ✓ | ✓ | ✓ | ✓ | ✓ | ✓ | ✓ | ✓ | ✓ | ✓ |
| Crime Lab Chemistry | ✓ | ✓ | ✓ | | ✓ | ✓ | | | ✓ | ✓ |
| Discovering Density | ✓ | ✓ | ✓ | ✓ | ✓ | ✓ | ✓ | | ✓ | ✓ |
| Earth, Moon, and Stars | ✓ | ✓ | | ✓ | | ✓ | | ✓ | ✓ | ✓ |
| Earthworms | ✓ | ✓ | ✓ | | | ✓ | ✓ | ✓ | ✓ | |
| Eggs Eggs Everywhere | | | ✓ | | | ✓ | | | ✓ | |
| Experimenting With Model Rockets | ✓ | ✓ | ✓ | ✓ | | ✓ | ✓ | | ✓ | ✓ |
| Fingerprinting | ✓ | ✓ | ✓ | | | | | | ✓ | |
| Frog Math:  Predict, Ponder, Play | ✓ | | ✓ | ✓ | | | | | ✓ | ✓ |
| Global Warming | ✓ | ✓ | ✓ | ✓ | ✓ | ✓ | ✓ | ✓ | ✓ | ✓ |
| Group Solutions | ✓ | ✓ | | ✓ | | ✓ | | | ✓ | ✓ |
| Height-O-Meters | ✓ | ✓ | | ✓ | | | | | | ✓ |
| Hide A Butterfly | ✓ | | ✓ | ✓ | | | | ✓ | ✓ | |
| Hot Water and Warm Homes | ✓ | ✓ | | ✓ | | ✓ | ✓ | | ✓ | ✓ |
| In All Probability | | | | ✓ | | ✓ | | | | |
| Investigating Artifacts | ✓ | | ✓ | ✓ | | ✓ | | | ✓ | ✓ |
| Involving Dissolving | ✓ | ✓ | ✓ | | ✓ | ✓ | | ✓ | ✓ | ✓ |

*This chart shows the major themes that permeate the GEMS guides.*

| | Systems and Interactions | Stability | Diversity and Unity | Models and Simulations | Matter | Patterns of Change | Energy | Evolution | Structure | Scale |
|---|---|---|---|---|---|---|---|---|---|---|
| Ladybugs | ✓ | | ✓ | ✓ | | ✓ | | ✓ | ✓ | ✓ |
| Learning About Learning | ✓ | | ✓ | ✓ | | ✓ | | ✓ | ✓ | ✓ |
| Liquid Explorations | ✓ | ✓ | ✓ | | ✓ | ✓ | | | ✓ | ✓ |
| Mapping Animal Movements | ✓ | ✓ | ✓ | | | ✓ | | ✓ | ✓ | |
| Mapping Fish Habitats | ✓ | ✓ | ✓ | | | ✓ | | ✓ | ✓ | |
| Math Around the World | ✓ | | ✓ | ✓ | | | | | | |
| Moons of Jupiter | ✓ | ✓ | | ✓ | | ✓ | ✓ | | ✓ | ✓ |
| More Than Magnifiers | ✓ | ✓ | | ✓ | ✓ | ✓ | ✓ | | ✓ | ✓ |
| Mystery Festival | ✓ | ✓ | ✓ | ✓ | ✓ | ✓ | | | ✓ | ✓ |
| Of Cabbages and Chemistry | ✓ | ✓ | | | ✓ | ✓ | | | | ✓ |
| On Sandy Shores | ✓ | | ✓ | | | ✓ | | ✓ | ✓ | ✓ |
| Oobleck: What Do Scientists Do? | ✓ | ✓ | | ✓ | ✓ | ✓ | | | ✓ | ✓ |
| Paper Towel Testing | ✓ | ✓ | | ✓ | ✓ | ✓ | | | ✓ | ✓ |
| Penguins And Their Young | ✓ | | ✓ | ✓ | | ✓ | ✓ | | ✓ | ✓ |
| QUADICE | ✓ | ✓ | | ✓ | | ✓ | | | | |
| River Cutters | ✓ | ✓ | ✓ | ✓ | ✓ | ✓ | ✓ | ✓ | ✓ | ✓ |
| Secret Formulas | ✓ | ✓ | | | ✓ | ✓ | | | | ✓ |
| Sifting Through Science | | ✓ | | | ✓ | | | | | ✓ |
| Stories in Stone | ✓ | | ✓ | ✓ | ✓ | ✓ | ✓ | ✓ | ✓ | ✓ |
| Terrarium Habitats | ✓ | | | | ✓ | ✓ | ✓ | ✓ | ✓ | |
| Tree Homes | ✓ | ✓ | ✓ | | | ✓ | ✓ | | ✓ | ✓ |
| Vitamin C Testing | ✓ | ✓ | | | ✓ | ✓ | | | | ✓ |
| The "Magic" of Electricity | ✓ | ✓ | | ✓ | ✓ | ✓ | ✓ | | | ✓ |
| Solids, Liquids, and Gases | ✓ | ✓ | ✓ | ✓ | ✓ | ✓ | ✓ | | ✓ | ✓ |
| Shapes, Loops & Images | ✓ | ✓ | ✓ | ✓ | | ✓ | | | ✓ | ✓ |
| The Wizard's Lab | ✓ | ✓ | ✓ | ✓ | ✓ | ✓ | ✓ | | | ✓ |

# Appendix C
# The Nature of Science

It is widely recognized that students are better able to gain a deep understanding of science if they learn what science is through their own direct experience; and then reflecting on how their personal experiences are related to what a scientist does. Further understanding comes from comparing their interactions with other students with the ways that scientific communities interact, and how science, as a body of knowledge and a way of finding out about the world, progresses.

Helping students (and teachers) gain an ever-increasing understanding of and appreciation for science is at the heart of the activity-based approach found in the GEMS series. In discussing our ideas about the nature of science and how the GEMS guides fit into this broad heading we came up with the following nine categories and have decided to list them as appropriate, along with themes, on the title page of each GEMS guides as they are revised, and on the title pages of new guides as they are published.

### Scientific Community

An individual lab team working alone cannot produce scientifically valid results, just as a single football team cannot play a football game by themselves. Science requires a community of people who can listen critically to each others' discoveries, build upon each others' ideas, test the results of experiments, and propose alternative theories. How the scientific community operates is illustrated in *Oobleck: What Do Scientists Do?* In their scientific convention, each lab group shares the results of their experiments, criticizes and improves each others' stated conclusions, and sometimes decides to return to the laboratory to verify results or explore new questions. There are numerous other examples that demonstrate the importance of a community of scientists throughout the GEMS series.

### Interdisciplinary

Virtually all fields of science require the application of a wide range of skills and knowledge, including not only mathematics, technology, and the natural sciences, but also social sciences, literature, language arts, history, drama, and art. Many GEMS guides illustrate the use of mathematics or technology in science and include "Going Further" activities that launch students into interdisciplinary extensions. Two of the guides, *Acid Rain* and *Global Warming and the Greenhouse Effect*, illustrate how environmental problems require the application of physics and chemistry, as well as mathematics, history, and the social sciences. *Investigating Artifacts* shows how archaeology, anthropology, and literature can provide different perspectives in understanding ancient peoples. In *Learning About Learning*, students combine the disciplines of psychology, neurology, anatomy, and ethics to understand thinking and learning.

## Cooperative Efforts

Unlike the common stereotype, scientists do **not** work alone. Most scientists work in laboratory groups, where several individuals can combine their skills and ideas. Cooperation between groups is helpful to improve scientific objectivity, or to approach phenomena from different points of view, or with different techniques. In most of the GEMS activities students usually work together in laboratory teams, ranging from two to five or six individuals. Individual students working within groups often are given specific roles to play, such as recorder, reporter, or timekeeper. In most GEMS units, the various laboratory groups pool their data or otherwise compare and contrast their results and insights. These important classroom techniques mirror the scientific process in many important ways. Perhaps the most cooperative of all the GEMS guides is *Group Solutions,* in which students work as teams to solve puzzles and problems in which each member of the team has a part of the solution. They can only solve the problems if they work together.

## Creativity and Constraints

Like artists, scientists are creative. They use their creativity to solve problems, plan experiments, or devise theories. The stories about the way creative energies and inspirations have opened up immense new vistas are legion. And yet, scientists must also work within constraints. Their creative ideas must be tested through experiments and subjected to the scrutiny of other scientists. Over its evolution, the scientific method of controlled experimentation has developed rigorous procedures that are recognized worldwide. *Paper Towel Testing* provides opportunities for students to be creative in devising experiments. *Earth, Moon, and Stars* invites students to imagine that they lived thousands of years ago, in the days when people believed the Earth to be flat, and to devise theories to explain how the sun moved from the west to the east during the night. These and other GEMS activities require that students fit their creativity within the constraints of evidence, observations, theories, and scientific methods.

## Theory-Based and Testable

Science is conducted within a theoretical framework that guides the scientists in deciding which hypotheses are important to test. For example, in the units *Mapping Animal Movements* and *Mapping Fish Habitats*, students observe and record the movements of animals, and then try to interpret their observations in terms of the ways that animal behaviors are adapted to their environments. Hypotheses about how a particular animal will behave in a particular environment are formulated as part of the broader theory of animal adaptation, and within the even broader framework of evolution. Students test these hypotheses as they conduct experiments with animals and their environments. Throughout the GEMS units, the emphasis on controlled experimentation and full analysis of results relates to the "theory-based and testable" nature of science. GEMS units that emphasize controlled experimentation include: *Hot Water and Warms Homes from Sunlight, Experimenting with Model Rockets, Vitamin C Testing,* and *Paper Towel Testing.*

### Changing Nature of Facts and Theories

A unique characteristic of science is that the body of knowledge that is widely held to be true today is very likely to change in the future. This is true even for many "facts," such as the number of moons around Jupiter, as well as for more general theories. Science units that are especially helpful in illustrating this characteristic are those in controversial areas. For example, *Global Warming and the Greenhouse Effect* explores a number of issues on which scientists do not agree: Is the world warming up? What will happen to the Earth's climate if the concentration of carbon dioxide and other "greenhouse" gases in the atmosphere continues to increase? The answers to all of these questions is changing on a daily basis as scientists learn more about the changing global environment.

### Evidence and Inference

How do we decide what facts and theories to believe? In brief, our current understanding results from inferences that are based on evidence. Just as a bloody fingerprint on a knife at a murder scene might provide evidence on which a jury could infer that a defendant is guilty, data can be used to support one theory in favor of another. Any kind of data can serve as evidence, as long as it was obtained by an unbiased observer. Just as in a court of law, the jury would not believe the evidence given by an unreliable witness, evidence must also pass tests of reliability. For example, in *Bubble-ology* students gather data to determine which soap solution makes the biggest bubble dome. In doing so, they face problems of reliability: "My hair fell into the bubble. Should I count it?" Resolving such questions is a very important aspect of science, since the nature of "truth" in science is no more than the inferences that we draw from evidence. The forensic science units, *Fingerprinting, Crime Lab Chemistry,* and *Mystery Festival* all provide interesting perspectives on evidence and inference in science, and the subtle ways that it differs from the evidence and inference in a court of law.

### The Importance of Questions

Scientists value good questions because the quality of the questions often determines the direction and usefulness of a research study. In the unit *Acid Rain,* students generate questions in the first activity, post the list of questions on the wall, and return to them later in the unit. Through this activity students see that questions are even more important than answers in driving scientific progress, and that answers change over time, as the students learn more about the subject. As mentioned in the GEMS Teacher's Guide, questions lead students to higher orders of thinking. Similarly, questions lead scientists to deeper levels of understanding.

### Objectivity and Ethics

Objectivity in science means making an effort to eliminate bias in testing hypotheses. In *Bubble-ology* the students design controlled experiments to ensure the brand of bubble solution they expect to create the biggest bubble will not receive an unfair advantage. This same issue of recognizing ones' own biases in science and always reporting the truth is vitally important in professional science, and in medical research, as past and current controversies continue to demonstrate. The intersection of science and society, whether considering societal consequences of scientific breakthroughs, or the personal responsibility of the individual scientist, has taken on even greater weight in the atomic age. Aspects of this ethical dimension are explored directly by students in both *Acid Rain* and *Global Warming*, as they face questions about the complex environmental ethics of such human activities as deforestation, massive burning of fossil fuels, and atmospheric emissions of pollutants that cause acid rain. In *Learning About Learning* students explore the ethical questions associated with research on humans, animals, and cell and tissue cultures.

## Real-Life Applications

Students like to know how classroom activities relate to the world that they experience every day, or to the world of professional science and technology. Students of any age understand, remember, and use abstract concepts far better when they discover their relevance in real life. In *Paper Towel Testing*, students apply scientific methods in trying to find the best buy in paper towels. In *Mapping Animal Movements*, they learn how techniques that they use in the classroom are like those used by professional field biologists. In Mystery Festival students conduct investigations just like forensic scientists. And in *River Cutters,* students use models and simulations to investigate how polluted groundwater migrates underground to river systems.

## Science and Technology

Science is aimed at asking questions and learning about the universe, with no practical applications necessarily expected. Technology, on the other hand, is a tool or technique that enables people to accomplish something more easily or effectively. Science and technology are interdependent. Developments in science often result in technological advances; new technologies are often used by scientists to make new discoveries. In *Oobleck: What Do Scientists Do?* students first act as scientists to learn all they can about Oobleck, a strange substance said to come from another planet. Then they act as technologists as they design spacecraft to land on an entire ocean of Oobleck. In *Bubble-ology,* students experiment to find out why some devices will make bubbles and some will not, then, using that knowledge, they design a "bubble-maker" that they think will make large (or small) bubbles. A number of GEMS guides include follow-up suggestions relating to the use of computers as technological tools to analyze data or project trends.

# NATURE OF SCIENCE CATEGORIES

| Categories | Scientific Community | Science and Technology | Changing Facts & Theories | Creativity and Constraints | Theory-Based and Testable | Cooperative Efforts | Objectivity and Ethics | Real-Life Applications | Interdisciplinary |
|---|---|---|---|---|---|---|---|---|---|
| Acid Rain | ✓ | ✓ | ✓ | ✓ | ✓ | ✓ | ✓ | ✓ | ✓ |
| Animal Defenses | | | | ✓ | | | | ✓ | |
| Animals in Action | ✓ | | ✓ | ✓ | ✓ | ✓ | ✓ | ✓ | |
| Ant Homes Under the Ground | | | | | | ✓ | | ✓ | ✓ |
| Bubble Festival | ✓ | ✓ | | | ✓ | ✓ | | ✓ | ✓ |
| Bubble-ology | ✓ | ✓ | ✓ | ✓ | ✓ | ✓ | ✓ | ✓ | ✓ |
| Build It! Festival | | ✓ | | ✓ | | ✓ | | ✓ | ✓ |
| Buzzing A Hive | | | | ✓ | | ✓ | | ✓ | ✓ |
| Chemical Reactions | ✓ | ✓ | ✓ | ✓ | ✓ | ✓ | | ✓ | |
| Color Analyzers | ✓ | ✓ | ✓ | ✓ | ✓ | ✓ | | ✓ | ✓ |
| Convection: A Current Event | ✓ | ✓ | ✓ | ✓ | ✓ | ✓ | | ✓ | ✓ |
| Crime Lab Chemistry | ✓ | ✓ | ✓ | ✓ | ✓ | ✓ | ✓ | ✓ | ✓ |
| Discovering Density | ✓ | | | ✓ | ✓ | ✓ | | ✓ | ✓ |
| Earth, Moon, and Stars | ✓ | ✓ | ✓ | ✓ | ✓ | ✓ | | ✓ | ✓ |
| Earthworms | ✓ | | ✓ | | ✓ | ✓ | ✓ | ✓ | ✓ |
| Eggs Eggs Everywhere | | | | ✓ | | | | ✓ | ✓ |
| Experimenting With Model Rockets | ✓ | ✓ | ✓ | ✓ | ✓ | ✓ | | ✓ | ✓ |
| Fingerprinting | ✓ | ✓ | ✓ | ✓ | ✓ | ✓ | ✓ | ✓ | ✓ |
| Frog Math: Predict, Ponder, Play | | | | | ✓ | ✓ | | ✓ | ✓ |
| Global Warming | ✓ | ✓ | ✓ | ✓ | ✓ | ✓ | ✓ | ✓ | ✓ |
| Group Solutions/Group Solutions, Too! | | | | | | ✓ | | ✓ | |
| Height-O-Meters | ✓ | ✓ | | | ✓ | ✓ | | ✓ | ✓ |
| Hide A Butterfly | | | | ✓ | | ✓ | | ✓ | |
| Hot Water and Warm Homes | ✓ | ✓ | ✓ | ✓ | ✓ | ✓ | ✓ | ✓ | ✓ |
| In All Probability | | | | ✓ | ✓ | | | ✓ | ✓ |
| Investigating Artifacts | ✓ | | ✓ | ✓ | | ✓ | ✓ | ✓ | ✓ |
| Involving Dissolving | | | ✓ | ✓ | ✓ | ✓ | | ✓ | ✓ |

This chart shows how major categories that illustrate the nature of science are represented in the GEMS guides.

| | Scientific Community | Science and Technology | Changing Facts & Theories | Creativity and Constraints | Theory-Based and Testable | Cooperative Efforts | Objectivity and Ethics | Real-Life Applications | Interdisciplinary |
|---|---|---|---|---|---|---|---|---|---|
| Ladybugs | | | | | | ✓ | ✓ | ✓ | |
| Learning About Learning | ✓ | ✓ | ✓ | ✓ | ✓ | ✓ | ✓ | ✓ | |
| Liquid Explorations | ✓ | | | | | ✓ | | ✓ | ✓ |
| Mapping Animal Movements | ✓ | | ✓ | ✓ | ✓ | ✓ | ✓ | ✓ | |
| Mapping Fish Habitats | ✓ | | ✓ | ✓ | ✓ | ✓ | ✓ | ✓ | |
| Math Around the World | ✓ | ✓ | | ✓ | | ✓ | | ✓ | ✓ |
| Moons of Jupiter | ✓ | ✓ | ✓ | ✓ | ✓ | ✓ | | ✓ | ✓ |
| More Than Magnifiers | ✓ | ✓ | ✓ | | ✓ | ✓ | | ✓ | ✓ |
| Mystery Festival | ✓ | | ✓ | ✓ | ✓ | ✓ | ✓ | ✓ | |
| Of Cabbages and Chemistry | ✓ | ✓ | ✓ | ✓ | ✓ | ✓ | | ✓ | ✓ |
| On Sandy Shores | ✓ | ✓ | | ✓ | | ✓ | ✓ | ✓ | ✓ |
| Oobleck: What Do Scientists Do? | ✓ | ✓ | ✓ | ✓ | ✓ | ✓ | | ✓ | ✓ |
| Paper Towel Testing | ✓ | ✓ | ✓ | ✓ | ✓ | ✓ | ✓ | ✓ | |
| Penguins And Their Young | ✓ | | ✓ | | | ✓ | | ✓ | ✓ |
| QUADICE | | | | ✓ | ✓ | ✓ | | ✓ | |
| River Cutters | ✓ | ✓ | ✓ | ✓ | ✓ | ✓ | ✓ | ✓ | ✓ |
| Secret Formulas | | ✓ | | ✓ | ✓ | ✓ | | ✓ | ✓ |
| Sifting Through Science | | ✓ | | | | | | ✓ | |
| Stories in Stone | ✓ | ✓ | ✓ | | | ✓ | | ✓ | |
| Terrarium Habitats | | | | | | ✓ | | ✓ | ✓ |
| Tree Homes | | | | | | ✓ | | ✓ | ✓ |
| Vitamin C Testing | ✓ | | ✓ | | ✓ | ✓ | ✓ | ✓ | ✓ |
| The "Magic" of Electricity | | ✓ | ✓ | | ✓ | ✓ | | ✓ | ✓ |
| Solids, Liquids, and Gases | | ✓ | | ✓ | ✓ | ✓ | | ✓ | ✓ |
| Shapes, Loops & Images | | ✓ | | ✓ | ✓ | ✓ | | ✓ | ✓ |
| The Wizard's Lab | | ✓ | | ✓ | ✓ | ✓ | | ✓ | ✓ |

# A Summary of National Content Standards In GEMS

| | Science as Inquiry | Science and Technology | History and Nature of Science | Science in Personal and Social Perspectives | Earth and Space Science | Life Science | Physical Science |
|---|---|---|---|---|---|---|---|
| Acid Rain | ✓ | ✓ | ✓ | ✓ | ✓ | ✓ | ✓ |
| Animal Defenses | ✓ | ✓ | | | | ✓ | |
| Animals in Action | ✓ | ✓ | ✓ | | | ✓ | |
| Ant Homes Under the Ground | ✓ | ✓ | ✓ | ✓ | ✓ | ✓ | ✓ |
| Bubble Festival | ✓ | ✓ | | | | | ✓ |
| Bubble-ology | ✓ | ✓ | ✓ | | | | ✓ |
| Build It! Festival | ✓ | ✓ | | ✓ | | | ✓ |
| Buzzing A Hive | ✓ | ✓ | ✓ | | | ✓ | ✓ |
| Chemical Reactions | ✓ | | ✓ | | | | ✓ |
| Color Analyzers | ✓ | ✓ | ✓ | | ✓ | | ✓ |
| Convection: A Current Event | ✓ | ✓ | | | ✓ | | ✓ |
| Crime Lab Chemistry | ✓ | | ✓ | | | ✓ | ✓ |
| Discovering Density | ✓ | | ✓ | | | | ✓ |
| Earth, Moon, and Stars | ✓ | ✓ | ✓ | | ✓ | | ✓ |
| Earthworms | ✓ | ✓ | | | | ✓ | |
| Eggs Eggs Everywhere | ✓ | ✓ | | | | ✓ | ✓ |
| Experimenting With Model Rockets | ✓ | ✓ | | | ✓ | | ✓ |
| Fingerprinting | ✓ | ✓ | ✓ | ✓ | | ✓ | ✓ |
| Frog Math: Predict, Ponder, Play | ✓ | | | ✓ | | | |
| Global Warming | ✓ | ✓ | ✓ | ✓ | ✓ | ✓ | ✓ |
| Group Solutions | ✓ | | | ✓ | | | |
| Height-O-Meters | ✓ | ✓ | | | ✓ | | ✓ |
| Hide A Butterfly | ✓ | ✓ | | | | ✓ | |
| Hot Water and Warm Homes | ✓ | ✓ | | ✓ | ✓ | ✓ | ✓ |
| In All Probability | ✓ | | ✓ | ✓ | | | |
| Investigating Artifacts | ✓ | ✓ | ✓ | ✓ | ✓ | ✓ | ✓ |
| Involving Dissolving | ✓ | | | | | | ✓ |

Please Note: This chart is a **summary** of content-related National Standards reflected in GEMS units, based in part on a preliminary analysis by GEMS Texas Network Coordinator Karen Ostlund, with GEMS Associates Tracey Harros, Mimi Halferty, and Pat Jones. It is intended as an overview to show how the content in the guides can be related to the standards—**a check mark is not meant to imply that a standard is necessarily a major focus in the guide, but that it is represented.** As the chapter on curriculum sequences in this handbook suggests, presentation of a number of units over time, which construct unifying concepts and processes in a coherent manner, is the best way to assist students in fully understanding content.

| | Science as Inquiry | Science and Technology | History and Nature of Science | Science in Personal and Social Perspectives | Earth and Space Science | Life Science | Physical Science |
|---|---|---|---|---|---|---|---|
| Ladybugs | ✓ | | | ✓ | | ✓ | |
| Learning About Learning | ✓ | | ✓ | ✓ | | ✓ | |
| Liquid Explorations | ✓ | | | | | | ✓ |
| Mapping Animal Movements | ✓ | ✓ | ✓ | | | ✓ | |
| Mapping Fish Habitats | ✓ | | | ✓ | | ✓ | |
| Math Around the World | ✓ | | ✓ | ✓ | | | |
| Moons of Jupiter | ✓ | ✓ | ✓ | ✓ | ✓ | ✓ | ✓ |
| More Than Magnifiers | ✓ | ✓ | | | | | ✓ |
| Mystery Festival | ✓ | ✓ | ✓ | ✓ | | ✓ | ✓ |
| Of Cabbages and Chemistry | ✓ | ✓ | ✓ | | | | ✓ |
| On Sandy Shores | ✓ | | | ✓ | ✓ | ✓ | ✓ |
| Oobleck: What Do Scientists Do? | ✓ | ✓ | ✓ | | ✓ | | ✓ |
| Paper Towel Testing | ✓ | ✓ | ✓ | ✓ | | | ✓ |
| Penguins And Their Young | ✓ | ✓ | | | | ✓ | ✓ |
| QUADICE | ✓ | | | | | | |
| River Cutters | ✓ | ✓ | ✓ | ✓ | ✓ | | ✓ |
| Secret Formulas | ✓ | | ✓ | | | | ✓ |
| Sifting Through Science | ✓ | ✓ | | | | | ✓ |
| Stories in Stone | ✓ | ✓ | ✓ | | ✓ | | ✓ |
| Terrarium Habitats | ✓ | ✓ | | ✓ | ✓ | ✓ | ✓ |
| Tree Homes | ✓ | | | ✓ | | ✓ | |
| Vitamin C Testing | ✓ | ✓ | ✓ | ✓ | | | ✓ |
| The "Magic" of Electricity | ✓ | ✓ | ✓ | | | | ✓ |
| Solids, Liquids, and Gases | ✓ | ✓ | ✓ | | | | ✓ |
| Shapes, Loops & Images | ✓ | ✓ | ✓ | | | | ✓ |
| The Wizard's Lab | ✓ | ✓ | ✓ | | | | ✓ |

# Appendix E
# House of Science: Transparency Masters

Communicating the new reforms in science education is not easy. Yet teachers and science specialists are frequently asked to make presentations at School Board meetings, at programs for the PTA, and to other science teachers and school administrators.

We have found that one of the most interesting and compelling ways to present the ideas behind the current reform movement is through the use of the "House of Science" metaphor presented with transparencies. This Appendix includes transparency masters for:

| | |
|---|---|
| pages 101–104 | Overview of each of the major reform documents |
| page 105 | A Common Vision (Ideas common to all of the documents) |
| page 106 | Quote from Henri Poincaré about the nature of science |
| pages 107-110 | "House of Science" metaphor (layer these on top of each other) |
| page 111 | "House of Science" and the entire educational system |
| page 112 | Many Houses of Science |

The transparencies on pages 101–104 can be used as you present the ideas outlined in Chapter 2 of this book, outlining the major reform documents. "A Common Vision," on page 105, can be used to introduce the ideas that are common to all of the reform documents, as described in Chapter 3. The quote from Henri Poincaré, pointing out that science is not just a disorganized pile of stones, is used to introduce the "House of Science" metaphor, which is described in Chapter 4. The transparencies on pages 107-110 are intended to be layered on top of each other, building a "House of Science" as you present the ideas describe on each transparency.

Once you produce transparencies from the masters, you may want to enhance their impact with the use of a set of colored overhead marking pens, but it is not essential to do so. We hope you will find them useful, and we wish you luck in implementing science education reform in your school or district!

> *Note: In many GEMS workshop presentations and others we have heard of nationwide, presenters, instead of using the transparencies, have actually constructed a table-top model house to make the metaphor even more concrete!*

*The educational foundations of our society are presently being eroded by a rising tide of mediocrity that threatens our very future as a Nation and a people.*

**—A Nation At Risk  (1983, page 5)**

# Project 2061

*Science for All Americans; and*
*Benchmarks for Science Literacy*

1. **Nature of Science**

2. **Nature of Mathematics**

3. **Nature of Technology**

4. **Physical Setting**

5. **Living Environment**

6. **Human Organism**

7. **Human Society**

8. **Designed World**

9. **Mathematical World**

10. **Historical Perspectives**

11. **Common Themes**

12. **Habits of Mind**

## A Vision of Science Literacy

# Scope, Sequence, and Coordination (SS&C)

***Every science, every year, for every student, grades 6-12***

# National Science Education Standards

1. Introduction

2. Principles and Definitions

3. Science Teaching Standards

4. Professional Development Standards for Teachers

5. Assessment Standards

6. Science Content Standards

| | |
|---|---|
| K-4 | Unifying Concepts and Processes |
| 5-8 | Science as Inquiry |
| 9-12 | Physical Science |
| | Life Science |
| | Earth and Space Science |
| | Science & Technology |
| | Science in Personal & Social Perspectives |
| | History and Nature of Science |

7. Program Standards

8. System Standards

# A Common Vision
## for Science Education Reform in the 1990s

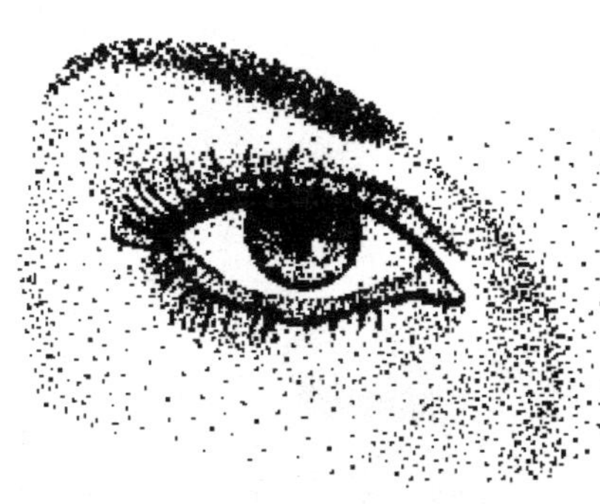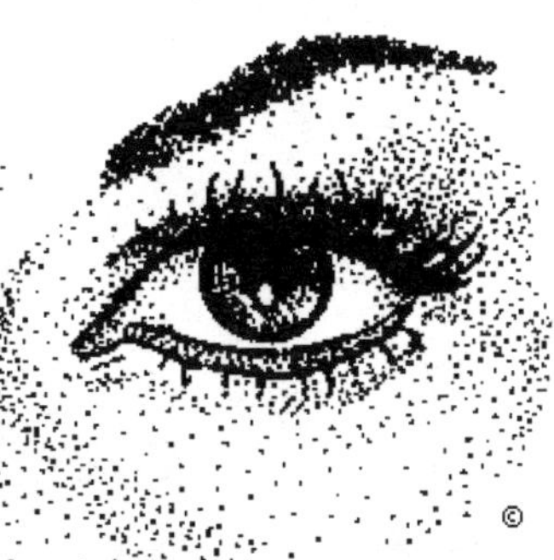

**Science for Whom?**

1. Science is for *all* students.

2. Strong commitment to diversity and equity.

**What Should Students Learn?**

3. Start with a foundation of inquiry skills.

4. Erect a framework of unifying concepts and processes.

5. Identify the most important concepts and theories.

6. Integrate ideas from different fields whenever possible.

7. Include examples illustrating the history and nature of science.

8. Show how science and technology inform societal issues.

**How Should Students be Taught?**

9. Students must construct their own understanding of science.

10. Students must collaborate in teams much of the time.

**What Must be Done to Implement the Vision?**

11. Align assessment with curriculum and teaching.

12. Mobilize the whole system to support the new reforms.

**Science is constructed of facts,
as a house is of stones.
But a collection of facts
is no more a science
than a heap of stones is a house.**

—Henri Poincaré

# Building a House of Science

## FOUNDATION:

### Inquiry Abilities and Understandings

- Asking Questions
- Conducting Investigations
- Using Appropriate Tools & Techniques
- Thinking Critically About Evidence & Explanations
- Constructing and Analyzing Alternative Explanations
- Communicating Scientific Arguments

### Science Process Skills
- Observing
- Communicating
- Comparing
- Ordering and Categorizing
- Relating
- Inferring
- Applying

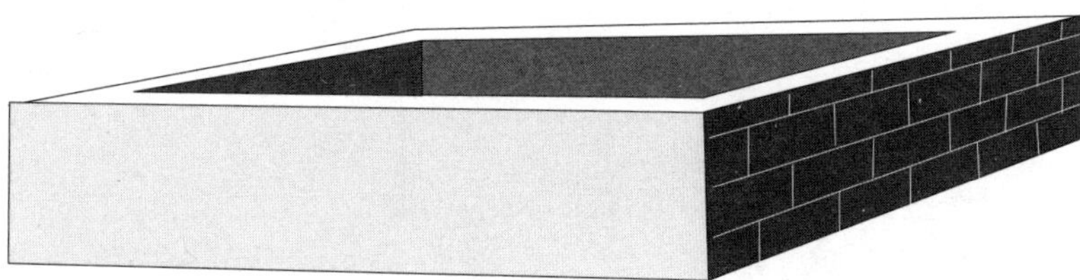

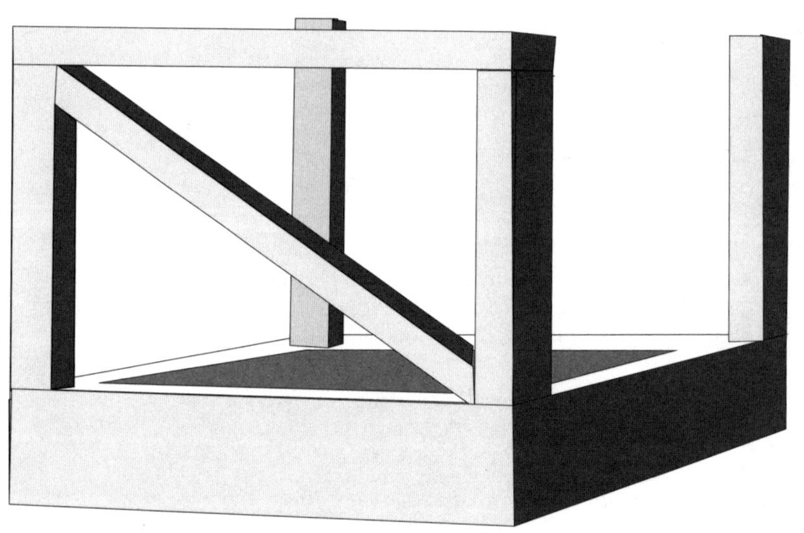

# Framework:

Evolution and equilibrium

Systems, order, and organization

Evidence, models, and explanation

Form and function

Constancy, change, and measurement

# Roof:

Science for *all* students. *The National Standards* emphasize that this embodies both "excellence and equity… regardless of age, gender, cultural or ethnic background, disabilities, aspirations, or interest and motivation in science."

# Mortar and Nails:
## Enjoyment & Curiosity

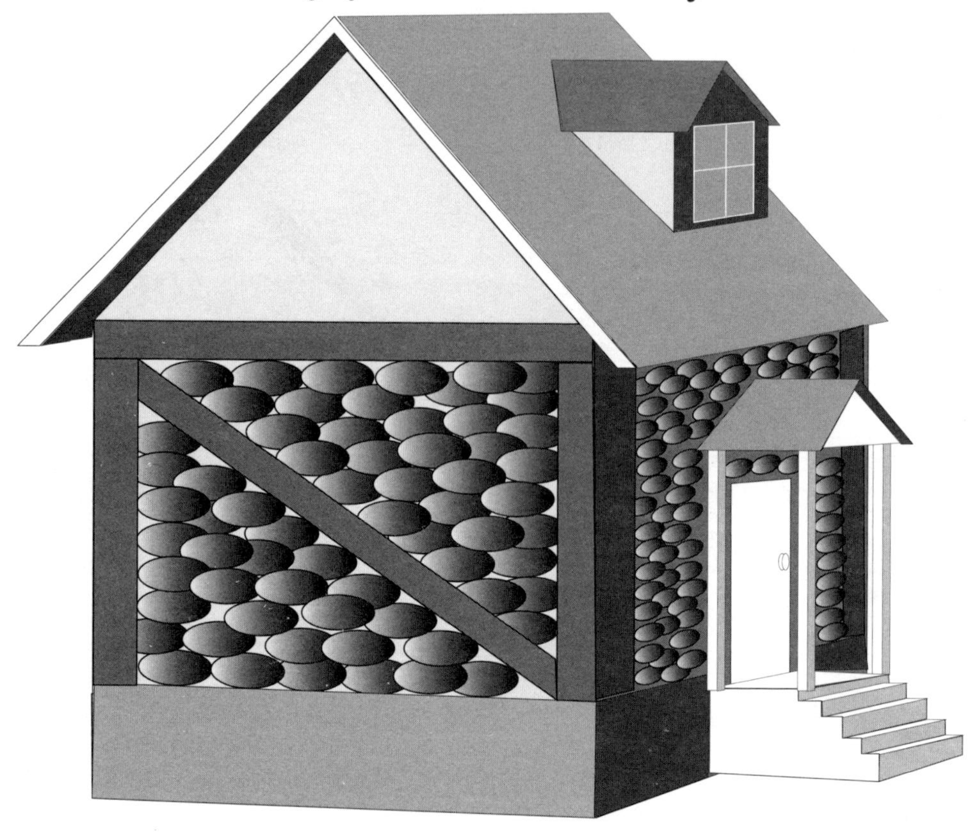

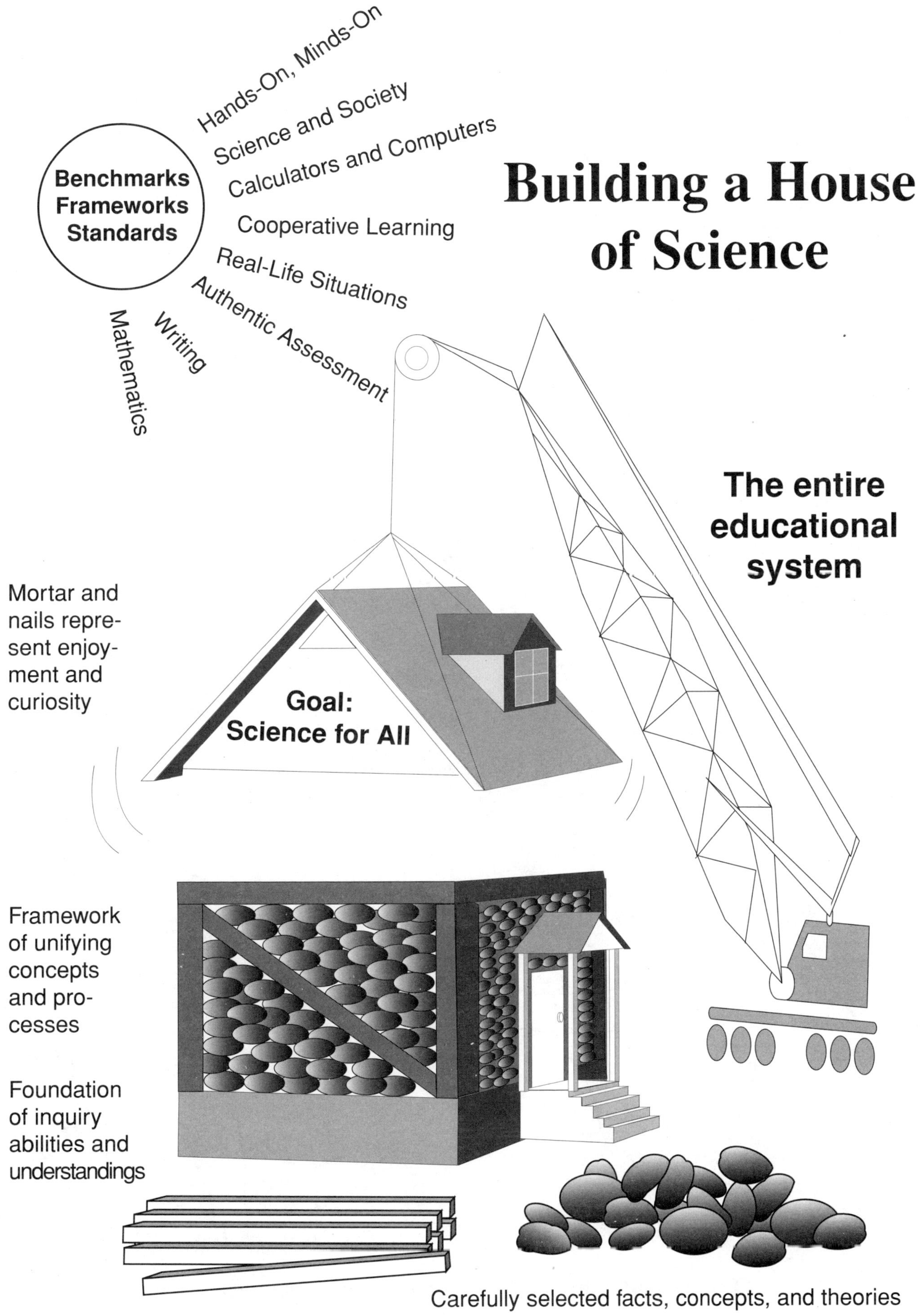

Benchmarks Frameworks Standards
Hands-On, Minds-On
Science and Society
Calculators and Computers
Cooperative Learning
Real-Life Situations
Authentic Assessment
Writing
Mathematics
Building a House of Science
The entire educational system
Mortar and nails represent enjoyment and curiosity
Goal: Science for All
Framework of unifying concepts and processes
Foundation of inquiry abilities and understandings
Carefully selected facts, concepts, and theories

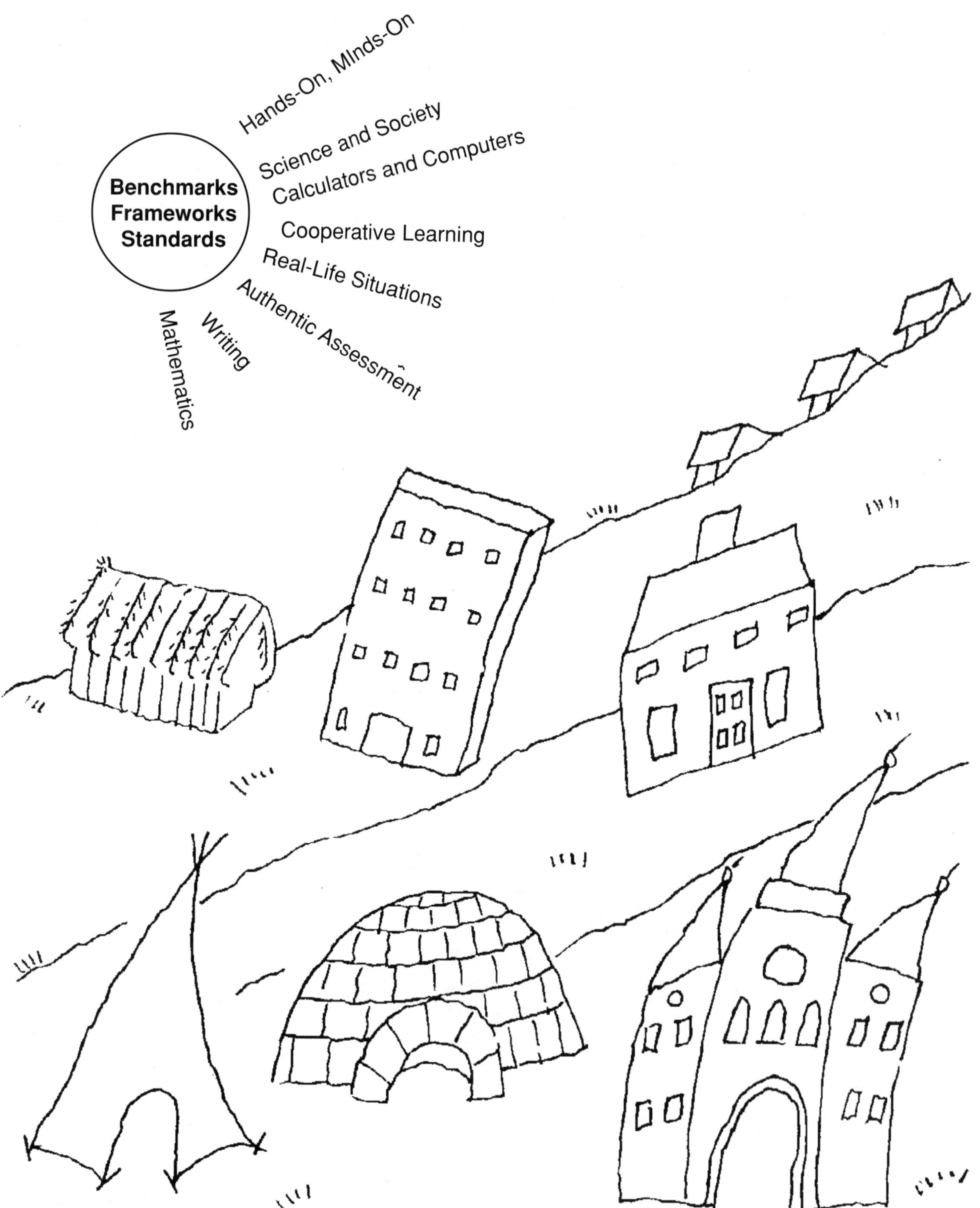


**Each child constructs her own house of science.**